Ethics in Technology

Ethics in Technology

A Philosophical Study

Topi Heikkerö

LEXINGTON BOOKS
Lanham • Boulder • New York • Toronto • Plymouth, UK

Published by Lexington Books
A wholly owned subsidiary of The Rowman & Littlefield Publishing Group, Inc.
4501 Forbes Boulevard, Suite 200, Lanham, Maryland 20706
www.rowman.com

10 Thornbury Road, Plymouth PL6 7PP, United Kingdom

British Library Cataloguing in Publication Information Available

Library of Congress Cataloging-in-Publication Data

Heikkerö, Topi, 1970–
Ethics in technology : a philosophical study / Topi Heikkerö.
p. cm.
Includes bibliographical references and index.
ISBN 978-0-7391-6864-6 (cloth : alk. paper)
1. Technology—Moral and ethical aspects. I. Title.
BJ59.H44 2012
174'.96—dc23
2012006069

The paper used in this publication meets the minimum requirements of American National Standard for Information Sciences Permanence of Paper for Printed Library Materials, ANSI/NISO Z39.48-1992.

Printed in the United States of America

For Thomas

Contents

Abbreviations	ix
Preface and Acknowledgments	xi
1 Introduction	1
Technology	4
Ethics	6
Approaches	7
Anticipations	10
Notes	12
2 The Question of Ethics in a Technological World	15
Challenges for Ethics	16
Applicability of Ethical Conceptions	26
The Meaning and Foundations of Ethics	29
How to Address the Questions	39
Notes	42
3 Hans Jonas: Technology and Responsibility	51
Hans Jonas and the Predicament of Modernity	52
Ethics and Ontology	58
Toward Updated Ethics	67
Critical Reflections	74
Conclusion	82
Notes	83
4 Albert Borgmann: Technology and the Good Life	93
On Science and Technology	95
Types of Discourse	98
Analysis of the Technological Culture	99

Approaches for a Reform 108
Assessing Borgmann's Project 115
Conclusion 124
Notes 125

5 Larry Hickman: Technology and Functionality 135
Pragmatism and the Philosophy of Technology 137
Hickman's Deweyan Philosophy of Technology 141
Pragmatism and the Ethics of Technology 154
Responsible Technology 159
Critique and Reflection 162
Conclusion 167
Notes 168

6 Thinking Further in Ethics and Technology 175
Being and Good 176
Toward an Ethics of Technology 182
Notes 196

7 Conclusions 201
Note 206

Bibliography 207

Index 225

Abbreviations

CPD	Albert Borgmann, *Crossing the Postmodern Divide* (1992)
HOR	Albert Borgmann, *Holding On to Reality* (1999)
IR	Hans Jonas, *The Imperative of Responsibility* (1979/1984)
JDPD	Larry A. Hickman, *John Dewey's Pragmatic Technology* (1990)
MM	Hans Jonas, *Mortality and Morality*, edited by Lawrence Vogel (1996)
PE	Hans Jonas, *Philosophical Essays* (1974)
PF	Albert Borgmann, *Power Failure* (2003)
PL	Hans Jonas, *The Phenomenon of Life* (1966/2001)
PPP	Larry A. Hickman, *Pragmatism as Post-Postmodernism* (2007)
PTTC	Larry A. Hickman, *Philosophical Tools for Technological Culture* (2001)
RAE	Albert Borgmann, *Real American Ethics* (2006)

TCCL	Albert Borgmann, *Technology and the Character of Contemporary Life* (1984)

Preface and Acknowledgments

The originating impulse for this book took place in summer 1999. I was working in the Group for Technology Studies at the Technical Research Centre of Finland. We were writing the report *Values and Participation in Technology Assessment*, which included passages on the questions concerning ethical evaluation of technology. These questions appeared to have direct bearing on our lives, but they were painfully difficult to address in a manner that would be academically, politically, and conscientiously sound. This experience led me to redirect my interest from more general ethical and philosophical concerns to philosophy of technology when I began to work on my doctoral studies. In brief, my rationale was that technology is both a public and a deep existential concern for all of us: one cannot escape technological shaping of our world, and yet this shaping evades rational analysis and assessment. This book is an evolved version of what once was my dissertation.

In working my way into philosophy of technology I was extraordinarily fortunate to share Carl Mitcham's help, mentorship, and friendship. Carl is a serious philosopher—classical style—in the frontier of cutting-edge thinking on technology.

Even a person who is slow to learn and reluctant to hear good advice needs a number of good people to help him write a book. Study groups, seminars, informal conversations over shared meals or coffee cups, walks, editing help, fierce and gentle criticism, and encouragement are some features of the philosophical friendship underlying this project. Thus, in a variety of ways, I am indebted to among others the following people: Kai Alhanen, Jussi Backman, Juan Bautista Bengoetxea, Albert Borgmann, Karin Christiansen, John Cornell, Erik Fisher, Jaana Hallamaa, Larry A. Hickman, Mikko Jauho, Tomi Külakoski, Heikki Kirjavainen, Timo Koistinen, Janne Kylliäinen, Philip LeCuyer, Jon Leydens, Susanna Lindberg, Juan Lucena,

Marc Lucht, David McDonald, Janne Nikkinen, Eeva Nyrövaara, Ville Päivänsalo, Craig Perfect, Tuukka Perhoniemi, Sami Pihlström, Mikko Rask, Hanne-Maaria Rentola, Kristin Shrader-Frechette, Jay Smith, Wha-Chul Son, David Strong, Caleb Thompson, Olli-Pekka Vainio, Markku Valtanen, Aku Visala, Llyd Wells, Langdon Winner, Russell Winslow, Sandy Woodson, Scott Yodel, and Henk Zandvoort.

The research project that enabled the writing of this book received funding from many sources. The University of Helsinki provided funding in various forms. The Academy of Finland, the Finnish Cultural Foundation, the Jenny and Antti Wihuri Foundation, and the Finnish Graduate School of Theology generously supported the project as well.

Very special thanks are due to a Lexington Books anonymous reviewer of the manuscript. This person did a more meticulous and generous job than I could have imagined. The improvements that resulted from the reviewer's even-handed comments are so many that I have not indicated them case by case. Finally, I want to thank Jana Hodges-Kluck, Kellie Hagan, and Eric Wrona from Lexington Books, who assisted this work into production.

In my private life, the support from my wife, Monika, and the joy of life presented by my son, Thomas, have been invaluable in the writing process.

Earlier versions of some sections have appeared in other publications, as follows: "The Good Life in a Technological World: Focal Things and Practices in the West and in Japan" in *Technology in Society* 27 (2005, No. 2), 251–259; "Ethics of Technology and the Notion of Nature: Hans Jonas's Philosophy of Life as a Grounding for His Ethics of Responsibility" in *Humanity at the Turning Point: Rethinking Nature, Culture, and Freedom. Essays on Contemporary Philosophy*, edited by Sonja Servomaa (Helsinki: Renvall Publications 23), 2006, 186–200; and "Material Culture and the Conditions for the Good Life: Albert Borgmann on Thinking about Ethics in Technological Culture" in *Praxiology and the Philosophy of Technology: The International Annual of Practical Philosophy and Methodology* 15, edited by Timo Airaksinen and Wojciech W. Gasparski (New Brunswick and London: Transaction Publishers), 2008, 215–229. I thank the previous publishers for the permission to use the material. Finally, special thanks are due to Hannu Raittila, for giving permission to use an excerpt from his novel in the beginning of chapter 5, and Andrew Chesterman, who let me use his translation of the text.

Chapter One

Introduction

Technology plays an extraordinary role in the contemporary world. Life in advanced societies depends on technology. To some extent human life has obviously always been dependent on technology. Prehistoric humans could not have survived without clothing and tools. But the extent of the current dependence is so much greater now than two or three centuries ago as to constitute a qualitative difference. This dependence penetrates physical, psychological, social, cultural, and all other layers and facets of modern human life. Moreover, technology itself has changed in character, and contemporary civilization takes considerable pride in the technologies it has created and utilizes. Technological innovations are regarded as the greatest achievements of humankind. The Internet, for instance, is considered a modern world wonder. However, spiritually or intellectually, technology is seldom addressed. Although many technological devices create some kind of acoustic noise either in their construction or in their operation, technology as a cultural phenomenon is most often surrounded by silence. In this sense, technology is situated in an intellectual blind spot: it forms the culturally invisible ground for most of our actions. It is commonly thought that "we just use technology, and there is not much more to it." Technological devices and systems are still most often perceived as transparent, neutral tools. Our culture continues to attribute the important questions solely to political decisions or the moral choices of individuals. One could, however, ask how large-scale a political decision would have to be in order to affect people's lives more than, say, television did by taking over the living room in the second half of the 20th century. Still, the common answer would be that the political and moral content of TV programs affects human lives but the device itself is transparent.

Although technology appears to be a blind spot of institutionalized intellectual reflection, some philosophers and cultural critics have attempted to capture the features of technology as a cultural and existential force. The term "philosophy of technology" (*Philosophie der Technik*) was first used by the left-wing Hegelian Ernst Kapp (1808–1896). Although Karl Marx, another left-wing Hegelian, did not explicitly address technology much, his theory of historical change rests on strong implicit assumptions concerning the role of technology: "The handmill gives you society with the feudal lord; the steam-mill with the industrial capitalist."[1] In the early 20th century, thinkers such as Max Scheler, Oswald Spengler, and José Ortega y Gasset addressed the question of how technology is going to affect human life and culture in general. This question continued to be treated by Lewis Mumford, Jacques Ellul, Martin Heidegger, Herbert Marcuse, and Ivan Illich, among others. The American pragmatist tradition has emphasized the instrumental, productive aspects of human existence and knowing. Consequently, pragmatism, especially John Dewey's thinking, encourages reflections on technology.[2]

Moreover, another beginning for reflection exists as well, one that takes off more from empirical reality than the philosophical armchair. Rachel Carson's *Silent Spring* (1962) and Thomas Kuhn's *The Structure of Scientific Revolutions* (1962) marked the beginning of science, technology, and society (STS) studies—if one wants to put a finger on a distinctive point. Arguably there was STS work even before: the work of Robert Merton, Jacob Bronowski, and C.P. Snow could be counted as such.[3] Subsequent STS, in any case, developed from an activist phase of the 1960s and early '70s into a professional academic field during the late 1970s and '80s. The social constructivist wave in STS has meant the introduction of anthropological fieldwork, careful case studies, and social scientific methods to science and technology studies. Bruno Latour, Steve Woolgar, Michel Callon, John Law, Wiebe Bijker, Harry Collins, and Trevor Pinch are some of the central academics in this movement.[4]

Hence, there exist in our culture at least two ways to address technology: philosophically and social scientifically. A certain complex, however, persists in the discussions on technology within both research traditions. People who investigate or think about technology usually have a mission: their discipline or, even worse, the whole culture, is in deep slumber concerning the role of technology in human life. Wake up! Bruno Latour's approach is illustrative of much academic writing on technology:

> For twenty years or so, my friends and I have been studying these strange situations that the intellectual culture in which we live does not know how to categorize. For lack of better terms, we call ourselves sociologists, historians, economists, political scientists, philosophers or anthropologists. But to these

> venerable disciplinary labels we always add a qualifier: 'of science and technology'. . . . Whatever label we use, we are always attempting to retie the Gordian knot by crisscrossing, as often as we have to, the divide that separates exact knowledge and exercise of power—let us say nature and culture. Hybrids ourselves, installed lopsidedly within scientific institutions, half engineers half philosophers, '*tiers instruits*' (Serres, 1991) without having sought the role, we have chosen to follow the imbroglios wherever they take us. . . . Yet our work remains incomprehensible, because it is segmented into three components corresponding to our critics' habitual categories. They turn it into nature, politics or discourse.[5]

But why should anyone be interested in thinking about technology? In general, one could claim, for instance, based on observations akin to those above, that it is awkward if one of the most fundamental conditions of current human life escapes intellectual reflection or is ignored by it. The very idea behind philosophy is the attempt to understand reality. It appears contradictory to this idea to leave technology out of the scope of philosophy. Admittedly, understanding for understanding's sake can sometimes be a redundant or luxurious goal. Yet, in this study, I argue that technological projects face unavoidable ethical questions. These questions are unavoidable because we may fail to explain the world or even abstain from the pursuit of explanations, but we cannot stop living in it. And if the world is to a large extent molded and created by technology, and if action in this world is mediated by technology, and if the results of the action are new technological devices, and if ethics addresses questions of good and right in human action, then there are inevitable ethical issues in technological action. Moreover, the agency in human action is transformed in the process of technological modernization. Technological mediation, for instance, makes killing human beings easy insofar as destruction becomes remote and nearly unnoticeable. Humans can also participate in creating ecological disasters without being aware of the effects of their actions due to the vast causal, spatial, and temporal dimensions of the technological networks. The responsibility, moreover, becomes hard to attribute. This altered agency deserves to be examined. These are fairly general and abstract questions. Yet they take a more concrete, particular, and practical form in policy decisions concerning technology in various areas. Consider education, especially engineering education; design processes of technological apparatuses; political reflection on who has the right and ability to decide about technology; and science and technology policy—for example, funding decisions and legislation.

Starting from such reflections, in this book I will examine how to think about ethics in a technological world. How this examination unfolds will become clearer in the following chapter. Here I try only to indicate briefly what I mean by the words "technology" and "ethics." After that I will briefly sketch the direction and content of the inquiry that is to follow.

TECHNOLOGY

It has proved nearly impossible to define technology in a satisfying manner: either the definition is so broad that technology appears to encompass all human activities, such as science, language, religion, and art, or it is so narrow that it omits some relevant perspectives that *prima facie* should be included. In this book, defining technology is not my main goal. Therefore, a flexible, heuristic definition suffices to give orientation to the inquiry. In his introduction to *Encyclopedia of Science, Technology, and Ethics* (2005) Carl Mitcham defines technology as follows: "'Technology' refers primarily to the modern activities of making and using artifacts, especially in applied science, engineering, medicine, decision-making, and management."[6]

This short definition points out a crucial feature of technology under current conditions: science and technology coexist and are difficult to separate from each other. Logically such a division is still possible, but as social, cultural, historical phenomena, modern science and modern technology are difficult to separate. Technology is extensively dependent on scientific knowledge and research activities. Consider digital technology, nanotechnology, and biotechnology. Science, nonetheless, has long been dependent on technology, as well. Galileo had his telescopes, modern physics needs huge particle accelerators, and gene scientists make use of electron microscopes and complex computer programs. This mutual dependence of science and technology is characteristic of the modern age. Earlier techniques used to be dependent on the personal skills of artisans, and science was more observational and contemplative. The modern form of production and knowing is sometimes referred to as "technoscience," with "technoscientific" as an adjective.[7] In fact, the English word "technology" already combines "technique" (Gr. *tekhnē*—skill, art, artifice) and "logos" (Gr. word, speech, reason, account, computation, ratio).[8] This etymology hints at both productive skill and systematic research elements in technology. In the following pages, I will draw on this symbiotic character of modern science and technology and address it in more detail to the extent it is relevant to the main themes.

In *Thinking through Technology: The Path between Engineering and Philosophy* (1994), Mitcham sketches a framework for a philosophical analysis of technology and perceives in it (1) knowledge, (2) volition, (3) activity (a. making, b. using), and (4) object/artifact components.[9] Mitcham's framework can function as a good preliminary, heuristic tool. Yet I would like to supplement his account in some respects. A first perspective concerns emphasizing technological systems. Quite evidently it does not suffice to depict technology as separate devices or artifacts. For a long time the system character has been essential in technologies. Technological products often do not appear to be separate objects but form expansive networks. Consider aqueducts, sewer systems, electricity grids, traffic networks, or bureaucratic management structures. Furthermore, this aspect of technological advance becomes even more essential with various forms of "high tech," especially in different information networks: the personal computer, for instance, truly broke through only with the Internet.

A second viewpoint is related to the systems perspective but moves to a subtler, more abstract plane. It appears that novel ways of thinking can create new technologies, and vice versa: technological advance has effected transformations in thinking. Development workers often face this issue, as it is not enough to export a device or even a system to a developing country. If the necessary mind-set is missing, technology fails. The transformations in the mind-set and thinking often happen on a very general and deep level in culture and may even go unnoticed. We may think, for instance, that in every age there was some view of sexuality. This kind of thinking, however, is "epoch-centered"; before a certain phase in the modern age there was no such notion as sexuality—which is a combination of biological, medical, psychological, and social "facts" that fuse with subjective desires, seen from the perspective of science as sexual needs. In earlier centuries, when the Christian religion had its normative hegemony and sex was more closely tied to physical consequences and risks, our scientific and individualistic notion of sexuality would have been inconceivable. Thus the phenomena we now perceive as "sexuality" were earlier perceived within a different framework; "sexuality" had not yet been abstracted from its physical and moral contexts. The process of abstraction involved political changes and scientific research, in addition to technological inventions. Comprehending sexuality as something separately manageable, however, is a subtle example of technological appropriation of human life. It, moreover, appears that sexuality belongs to our certainties; it is more than a concept, theory, or a thought: it is something on which we base our thinking.

This theme could be called technology as cultural paradigm, or frame.[10] In 20th-century philosophy, Ludwig Wittgenstein's notion 'form of life' (*Lebensform*) refers to these kinds of certainties. Form of life in Wittgenstein refers to a set of pragmatic and intersubjective conditions in human life that give a language its meaning. Speaking a language is action; it is part of lived life and shared practices. From this it follows that a meaning of a particular thought or sentence rests on many other thoughts, engaged action, and certainties, many of which are not consciously intended or possible to weigh neutrally.[11] Now, I am suggesting that technoscientific advance has altered our 'certainties.'

In Martin Heidegger's thinking, 'enframing' (*Ge-stell*) names the framework within which Being is revealed during the technological epoch. Enframing refers to a way of disclosing the world. There is always such a way: in the Middle Ages, Being was unconcealed as creatures in relation to their

Creator; in the modern age, Being becomes unconcealed as a resource (*Bestand*) to be used. Within enframing, modern science and technology disclose a truth about the world, but another way of disclosing would open the world differently.[12]

This last remark also reveals that the word "world" in the formulated problem is pregnant. To speak about "technological culture" or "technological society" would be inadequate. Technology is a social phenomenon; it is a cultural phenomenon, but it is a material phenomenon as well. In a sense, it is also a natural phenomenon: technologies are developed in dialogue with knowledge about nature. In a limited sense it is true that technological devices often are applications of knowledge about nature. This is one reason why Latour speaks about technological phenomena as "hybrids." And as the ideas of Wittgenstein and Heidegger suggest, technological advance affects—and presumes—the ways in which we encounter the world, or even what world, or nature, we encounter.[13] This does not necessarily need to lead into "idealism" or relativism but more centrally shows that technology poses an ontological question too.

Thus, with these two additions—the systems perspective and the ontological dimension—I will employ Mitcham's characterization of technology as a heuristic tool. This outline will be specified and expanded upon in the next chapter.

ETHICS

Ethics I take to be a discourse that addresses questions of the good and the right. By this I mean to set off with no commitment to any particular ethical theory. I also wish to be explicit that the primary aim is at normative ethics. "How to think about ethics in a technological world" means thinking about ethics in a normative sense. Moreover, in a philosophical inquiry ethics means rational discourse on the matters of good and right. This, however, does not imply any partisan rationalism: if there are any ethical phenomena that fall outside the rational discourse, then they should be identified as such. Finally, the exact dividing line between ethics and politics can hardly be drawn. In the particular case of ethics and technology this is even more true, since technology is often a public project and concern. Therefore, ethics here means also ethics/politics. This study, however, concentrates on "the ethical side" of politics, when it touches upon politics: forms of social justice, rights to participation, community as an ethical-political forum, for instance, will be discussed, but political power questions, organization theories, or political economy will not be the focus.

In short, by ethical thinking I mean locating questions of good and right amidst contexts within and related to technology. As with technology, this sketching of ethics will gain some concreteness and detail in the following chapter. In what follows, the problematic relationship between ethics and technology will be mapped. Further, three promising approaches for addressing the problems will be reviewed. In a sense, this forms a survey of the current state of affairs in the ethics of technology. Reviewing the approaches naturally leads into critical analysis. Finally, through mapping the problems and discussion of the existing positions, thinking about ethics in technology further becomes possible.

APPROACHES

I will introduce the approaches of three contemporary thinkers who have created some of the most relevant ways to think about ethics in technological conditions. Although thinking about technology has been in the margins of academic research and philosophy, three approaches obviously do not exhaust the field. My judgment in the selection of the approaches is based on the following considerations. First, as I am focusing on ethics rather than politics I will not take the venerable tradition of Critical Theory and its heirs under scrutiny. The work of the critical theorists, in fact, has been better studied in its own right than philosophy of technology. Second, as the brief notes on the world-disclosing and constituting aspects of technology show, even an ethically oriented inquiry into technology has to attend to its ontological dimensions as well. I have therefore been interested in thinkers that are also strong in this respect, for the ontological questions are quite evidently often ignored by practically oriented ethical approaches to technology-dependent moral problems. Third, I have kept the number of thinkers small, as I prefer in-depth analysis of a few positions to a larger number of more superficial surveys. Finally, the three approaches present thinking that aims at transcending the boundaries of modern philosophical traditions, especially the divide between analytic and continental philosophy.

Hans Jonas (1903–1993) and Albert Borgmann come from Heideggerian tradition and can be regarded as phenomenologists, but not in the narrow, orthodox sense of the word. They both integrated elements from Anglo-American intellectual culture into their thinking. Jonas was well informed in natural sciences, especially life sciences, and attempted to overcome the gap between two cultures in his philosophical projects. He, moreover, had a crucial role in the development of the Hastings Center, a research institute in bioethics founded in 1969. Borgmann draws on Heidegger's legacy in a manner that attempts to be as sober and scientific as possible. Borgmann is a scientific realist and well informed in natural science. In social philosophy his work is in constant dialogue with John Rawls's project. Larry Hickman, the third thinker, is a Deweyan pragmatist. He, nevertheless, dialogues with other philosophical traditions extensively, albeit his position is clearly that of a pragmatist. Hickman perceives Deweyan pragmatism as a solution for most, if not all, problems of standard philosophy in general and philosophy of technology in particular.

The relevance of these three thinkers can be succinctly characterized as follows. Jonas's importance arises from his introducing 'responsibility,' which is a grounding notion for thinking about our duties to future generations.[14] The notion of responsibility also provides a sound philosophical grounding for the precautionary principle, a central tool in contemporary policy-making. Moreover, his analysis of nihilism inherent in the modern worldview sheds light on the effect that technoscientific rationality has had on ethics as a rational discourse. Finally, Jonas's important work deserves to be better known in Anglophone academia. Borgmann's position is important as it explicitly raises the question of technology and the good life. How does technological advance affect human excellence? This question is always present but most of the time kept implicit both in academic disciplines and the culture at large. Borgmann also further develops Heidegger's philosophy of technology through a phenomenological analysis of everyday life in technologized culture. This analysis provides understanding that is very relevant for contemporary ethics. Hickman's position earns its merit through creating a consistent philosophy of technology that truly manages to rethink various categories that have prevented an adequate understanding of technology. Hickman achieves his position drawing explicitly on John Dewey's legacy. This feature constitutes a second, external reason for selecting his work as a source: Hickman's contribution serves also as a representative of the recent pragmatist tendencies in philosophy of technology.[15] His general ethos, moreover, is "pro technology" in a qualified and enlightened manner, while Jonas's and Borgmann's ethos might appear quite critical of technology to many.

The main intellectual context for this book is academic philosophy of technology that has emerged during the last five decades or so. The themes of ethics and technology, however, have been addressed in applied ethics as well.[16] Engineering ethics, research ethics, and medical ethics, for instance, operate in technological contexts. In applied ethics, however, technology in itself is seldom analyzed, but the ethical issues are perceived as existing between or within human agents. To some extent I will review these applied ethics approaches. Moreover, as I already noted, social scientific studies of technology provide relevant perspectives for ethics discourse that addresses technology. As I aim at treating ethics and technology on a general level, the applied ethics perspective is not in the focus but supplementary in character. The same holds for STS studies: they serve as providers of empirical information for philosophical analysis.

As the above suggests, bits and pieces of the ethics of technology exist in the academic community today, but they are scattered across multiple disciplines. Since the ethics of technology as an integral discipline does not yet exist, I begin by discussing the themes of ethics *and* technology. My ambition, nevertheless, is to look for relevant possibilities for combining disciplinary perspectives and work toward the ethics *of* technology. This objective sets this inquiry on an interdisciplinary trajectory.

Obviously, earlier attempts to advance ethics of technology exist. Efforts to cover the territory have recently been made from several directions previously mentioned: from philosophy of technology, from applied ethics, and from STS. Philosophers of technology in addition to those analyzed in this study have engaged in these efforts. Langdon Winner, Andrew Feenberg, and Carl Mitcham, for instance, have developed important approaches. Winner, in *Autonomous Technology: Technics-out-of-Control as a Theme in Political Thought* (1977), *The Whale and the Reactor: A Search for Limits in an Age of High Technology* (1986), and numerous articles, has analyzed technology as a political phenomenon. Feenberg, for example, in *Critical Theory of Technology* (1991) and *Questioning Technology* (1999), has drawn on the legacy of Critical Theory, especially Herbert Marcuse's, and developed further the ideas of critical reason and democratization of technology. He has, for instance, created more nuanced views of technological change with the help of social constructivist technology studies. Mitcham has worked toward the ethics of technology in his studies on the systematic philosophy of technology and on the history of the field. He has called his *Thinking through Technology* (1994) "a prolegomena for ethics of technology." *Encyclopedia of Science, Technology, and Ethics* (2005), edited by Mitcham, is in many respects a summa of the current knowledge and thinking in the field. Paul B. Thompson's work on the philosophy of agriculture and Andrew Light's on environmental philosophy in many senses come close to questions raised in this study. Of the younger generation, the works of Tsjalling Swierstra, Joseph Keulartz, Peter-Paul Verbeek, and Wha-Chul Son in different respects address the same issues as this study. Swierstra and Keulartz have attempted to develop a flexible pragmatist approach to the ethics of technology. Verbeek has assessed ways to think about 'things' ethically. Son has treated the problematic relationship of modern technology and democracy.[17]

ANTICIPATIONS

Against the background thus indicated, let me briefly outline the course of this inquiry. In the second chapter, I will analyze the relationship of ethical thinking and modern technology in more depth. I begin by considering the challenges technological advance has posed for ethics: new areas of human enterprise, expanded causal and temporal dimensions, risks, uncertainties, and vulnerabilities. I continue by reviewing briefly how these challenges have been addressed in such applied ethics fields as engineering ethics, bioethics, environmental ethics, computer ethics, research ethics, and business ethics. Then I look at wider social and political issues including just distribution of cost, benefits, harms, and risks in technologized societies. In a second major section of the chapter, I consider the applicability of traditional ethical frameworks in the altered context of the technological world. A third section addresses the questions concerning the foundation of ethics. In the 20th century, thinkers in both triumphant and lamenting tones put forward an idea that ethics has become irrelevant, meaningless, or lost its foundation in the technoscientific modernization process.

The third chapter consists of Hans Jonas's contributions to the ethics of technology. Jonas was one of Heidegger's influential Jewish students, who in his academic work addressed a wide variety of topics, from ancient religions and post-Holocaust theology to the problems of bioethics, such as human cloning. The chapter will open with an account of Jonas's analysis of what he perceived as the nihilistic predicament of the modern Western culture: he thought the worldview in the modern age, due to its dualistic underpinnings, was unable to provide a framework for serious thinking about morality. This connects to his phenomenological analysis of organic life, since Jonas argued that an appropriate understanding of life is able to correct some of the dualistic-nihilistic defects. Organic life forms "an ontological grounding for ethics" in Jonas's thinking. The treatment of these first two topics is mostly based on his collections of essays *The Phenomenon of Life: Toward a Philosophical Biology* (1966) and *Philosophical Essays: From Ancient Creed to Technological Man* (1974).

A third topic in this chapter is Jonas's constructive contribution to the issue: an ethics of responsibility that he presents in *The Imperative of Responsibility: In Search of an Ethics for the Technological Age* (German 1979, English 1984). The imperative, which is introduced in the book, itself has many formulations, such as: "Act so that the effects of your actions are compatible with the permanence of genuine human life." "Do not compromise the conditions [necessary] for the indefinite continuation of humanity on earth." "In your present choice, include the future wholeness of Man [*sic*] among the objects of your will."[18] I will address the ethics of responsibility and its foundations and corollaries that are relevant today.

In the fourth chapter I introduce and address Albert Borgmann's thinking. Borgmann is a North American (native German) philosopher who has analyzed technology as a social, ethical, and political phenomenon within the past thirty years. The chapter will open with Borgmann's analysis of technology as a way to be in the world in contemporary culture. The technological "pattern to take up with reality" he names the device paradigm. Borgmann argues that the device paradigm is useful when it is in its proper place in human life. In contemporary culture, however, it has extended too far and often appears to be the only way in which reality is encountered. In the second section I go on to introduce Borgmann's remedies for the culture that knows only a technological way to be in the world. The starting point for Borgmann's reform is the discovery of original, non-technological things and practices amidst the device way of thinking. The beginning of this exploration is somewhat poetic and ontological, but it expands to discussions of ethics and politics as well: how is authentic, "focal" life possible in technological society? The third section consists of a critical assessment of Borgmann's approach and temporary conclusion concerning his work.

The central presentation of Borgmann's philosophy of technology is *Technology and the Character of Contemporary Life: A Philosophical Inquiry*, published in 1984. *Crossing the Postmodern Divide* (1992) presents the arguments of his earlier, more scholarly book in a more accessible manner. *Holding On to Reality: The Nature of Information at the Turn of the Millennium* (1999) analyzes information technology, which is in the process of creating alternative realities. *Power Failure: Christianity in the Culture of Technology* (2003) takes a look at Christian faith in the technological world. Borgmann's most recent book, *Real American Ethics: Taking Responsibility for Our Country* (2006), presents his take on the American way of life and the explicit treatment of moral philosophy. Ethics, nevertheless, is at least indirectly present in Borgmann's work all the time.

In the fifth chapter I introduce and discuss Larry Hickman's philosophy of technology. Hickman is a Dewey scholar who has created a consistent philosophy of technology that provides him a framework to address social and political issues from a progressive perspective. The first part of the chapter consists of Hickman's general approach to philosophy of technology, his Deweyan productive pragmatism. His key ideas include a recasting of the praxis-theory distinction, anti-foundationalism, anti-essentialism, a broad definition of technology, emphasis on problem solving and experimentation, knowing as a technological artifact, and "publics" as products. Using these ideas, Hickman develops a perspective that seeks to overcome the limitations of traditional Western philosophy in general and classical philosophy of technology in particular. Although Hickman draws heavily on Dewey, his position is still an independent account in philosophy of technology. In the second section I look at how Hickman's general philosophy of technology tackles ethics. Hickman does not explicitly address the ethics of technology. Yet ethics is implicitly present in his thinking much of the time. Hence the main contribution of this section is to distill Hickman's implicit ethics and show the ethical implications of his work. By and large they have to do with reflecting on our technological practices, education, and democracy. The third section consists of a critical assessment of Hickman's contribution to the themes of the study.

The central volumes in which Hickman presents his philosophy of technology are *John Dewey's Pragmatic Technology* (1990), *Philosophical Tools for Technological Culture: Putting Pragmatism to Work* (2001), and *Pragmatism as Post-Postmodernism: Lessons from John Dewey* (2007). In the 2001 and 2007 books he forms more independent positions and answers to the critique his first book received from philosophers of technology. Hickman has been active in discussing others' work as well. I will especially introduce Hickman's criticisms of Borgmann's ideas in chapter 4.

In the sixth chapter I develop further the themes of the preceding chapters. I place ethics of technology in the context of the larger problem of the modern project. I revisit the question of facts and values severally encountered in earlier chapters. I sketch both "a minimum task for ethics of technology" and "a maximum ideal" for what can be expected from it. I centrally argue for the relevance of mid-level ethics in technological contexts. By "mid-level" I mean ethical questions that exist between the individual-based micro-level and broad discussions on the principles of justice. The seventh chapter consists of the conclusions and restatement of the central arguments of the book.

NOTES

1. Karl Marx, *The Poverty of Philosophy* (2005), 112.
2. Carl Mitcham, *Thinking through Technology: The Path between Engineering and Philosophy* (1994), 20–24, 45–49, 78–88, 247–248. Carl Mitcham and Helen Nissenbaum, "Technology and Ethics," *Routledge Encyclopedia of Philosophy* (1998).
3. See, e.g., Robert K. Merton, *The Sociology of Science* (1973); Jacob Bronowski, *Science and Human Values* (1990); and C. P. Snow, *The Two Cultures* (1993).
4. Stephen Cutcliffe, "Science, Technology, and Society Studies," *Encyclopedia of Science, Technology, and Ethics* (2005). Wiebe Bijker, "Understanding Technological Culture through a Constructivist View of Science, Technology, and Society," *Visions of STS: Counterpoints in Science, Technology, and Society Studies* (2001).

5. Bruno Latour, *We Have Never Been Modern* (1993), 3. "Hybrid" and "imbroglio" in Latour's work refer to things that cannot be classified in the categories of "nature" and "society" without problems. Technological devices and scientific instruments, for instance, are in Latour's reading "hybrids." Latour refers to Michel Serres's book *Le tiers instruit* (Paris: Bourin, 1991).

6. Mitcham, Introduction, *Encyclopedia of Science, Technology, and Ethics* (2005), xiii.

7. Larry A. Hickman, *Philosophical Tools for Technological Culture* (2001), 42–43. Gilbert Hottois ("Technoscience," *Encyclopedia of Science, Technology, and Ethics*, 2005) defines it as follows: "Technoscience refers to the strong interactions in contemporary scientific research and development (R&D) between that which traditionally was separated into science (theoretical) and technology (practical), especially by philosophers."

8. George Henry Liddell and Robert Scott, *A Greek-English Lexicon* (1983).

9. Mitcham, *Thinking through Technology*, 154–160.

10. Ibid., 154–160.

11. Ludwig Wittgenstein, *Philosophical Investigations* (1997), §19ff; *On Certainty* (2004), §141, §410, §611. "The child learns to believe a host of things. I.e. it learns to act according to these beliefs. Bit by bit there forms a system of what is believed, and in that system some things stand unshakably fast and some are more or less liable to shift. What stands fast does so, not because it is intrinsically obvious or convincing; it is rather held fast by what lies around it" (*On Certainty*, §144). Wittgenstein's remarks on technoscientific civilization: *Culture and Value*, 6–7, 14, 19, 56, 63.

12. Martin Heidegger, *Discourse on Thinking* (1966), 43–57; *What Is Called Thinking?* (1966), 3–19, 229–244; *The Question Concerning Technology and Other Essays* (1977), 3–35, 115–154. Hubert Dreyfus, "Heidegger on the Connection between Nihilism, Art, Technology, and Politics," *The Cambridge Companion to Heidegger* (1993). Mitcham, *Thinking through Technology*, 49–57. Graham Harman, *Heidegger Explained* (2007), 127–141.

13. Robert B. Pippin, "On the Notion of Technology as Ideology," *Technology and the Politics of Knowledge* (1995). Lawrence Vogel, "Hans Jonas's Diagnosis of Nihilism: The Case of Heidegger," *International Journal of Philosophical Studies* (1995); "Does Environmental Ethics Need a Metaphysical Grounding?" *Hastings Center Report* (1995).

14. Deborah G. Johnson, "Responsibility: Anglo-American Perspectives"; Hans Lenk, "Responsibility: German Perspectives"; Mitcham, "Responsibility: Overview," *Encyclopedia of Science, Technology, and Ethics* (2005).

15. Mitcham, *Thinking through Technology*, 71–78. Don Ihde, "Has the Philosophy of Technology Arrived? A State-of-the-Art Review," *Philosophy of Science* (2004). Jozef Keulartz, Maartje Schermer, Michiel Korthals, and Tsjalling Swierstra, "Ethics in Technological Culture: A Programmatic Proposal for a Pragmatist Approach," *Science, Technology, and Human Values* (2004). Peter-Paul Verbeek, *What Things Do* (2005), 99–119.

16. Brenda Almond, "Introduction: Ethical Theory and Ethical Practice," *Introducing Applied Ethics* (1995). Mitcham and Nissenbaum, "Technology and Ethics." Ruth Chadwick, "Applied Ethics," *Encyclopedia of Science, Technology, and Ethics* (2005).

17. Paul T. Thompson, *The Spirit of the Soil: Agriculture and Environmental Ethics* (1995). Andrew Light, "Environmental Ethics," *A Companion to Applied Ethics* (2003). Keulartz et al., "Ethics in Technological Culture." Tsjalling Swierstra and Jaap Jelsma, "Responsibility without Moralism in Technoscientific Design Practice," *Science, Technology, and Human Values* (2006). Verbeek, *What Things Do*. Wha-Chul Son, "Philosophy of Technology and Macroethics in Engineering," *Science and Engineering Ethics* (2008).

18. Hans Jonas, *The Imperative of Responsibility: In Search of an Ethics for the Technological Age* (1984), 11.

Chapter Two

The Question of Ethics in a Technological World

Studies on the Second World War suggest that individual solders killed surprisingly few enemies with their personal weapons. Killing with group weapons, such as machine guns, tanks, launchers, artilleries, and planes, was far more common. In this phenomenon at least two factors are at work: the patterns of group action and technological mediation. In earlier eras, the first was used more in military praxes; with severe discipline, solders were trained to overcome their personal aversion to killing. With modern technology, the role of mediated killing has intensified in warfare. In fact, it appears that the more an act of killing is technologically mediated, the easier it is for a human being to kill. Thus technological advances have transformed the character and boundaries of killing. This raises philosophical questions concerning morality, human agency, and technology that intertwine in the conditions of war. Questions such as these exemplify the requirements for rethinking ethics in technological contexts.[1]

To introduce those requirements in this chapter, I frame and analyze the relationship between ethics and technology. I perceive this relationship as having three distinct "facets." The first major section looks at the challenges that technological advance has posed to ethics. This section also includes a brief review of academic responses to these challenges. The second section examines the claim that traditional ethical concepts, theories, and frameworks lack adequacy or applicability in technologized conditions. Concepts such as harm and right, frameworks such as distinctions between making (*poiesis*) and doing (*praxis*) will be discussed. The third section raises a question concerning the foundations of ethics. Philosophers have triumphantly and lamentably argued that modern technoscientific advance makes ethics unnecessary. I review these claims with a characterization of the somewhat dire situation of ethics in 20th-century philosophy. The last section sketches an interdisciplinary viewpoint for orientation to the questions of ethics and technology.[2]

CHALLENGES FOR ETHICS

Tangible problems have followed the technological modernization process. They became dramatically visible at times of accidents such as Three Mile Island, Bhopal, and Chernobyl[3] and when authors revealed the previously unknown dangerous character of asbestos, DDT,[4] CFCs,[5] and X-rays. Slightly less dramatic but more serious is the permanently present ecological predicament. Thinning of the ozone layer, climate change, pollution, rapid extinction of species, and erosion, to name some central aspects of the crisis, are threatening all life on earth. Furthermore, pollution causes human suffering in the form of diseases.

Strictly speaking, not all ecological, human, or global problems are ethical problems. Obviously, a host of ethical questions exists in relation to the above problems of humankind. But ethical questions are questions concerning good and right in the context of human action. Ecological imbalance is merely an ecological issue until it is set in relation to human agency and seen in the light of normative concepts.

Regional Approaches

An apparent, "empirical" way to determine what ethics and ethical questions in the context of technology mean is to look at some new branches of ethics that have emerged to address the challenges. These are professional ethics for engineers, computer ethics,[6] biomedical ethics, environmental ethics, nuclear ethics,[7] agriculture and food ethics,[8] research ethics,[9] and some forms of business ethics.[10] In these disciplines, ethics is being more or less successfully regionalized into distinct academic fields with defined issues.[11] I will briefly describe engineering ethics, environmental ethics, and bioethics as examples of applied ethics tackling the challenges of technology.

Professional ethics for engineers, or engineering ethics, has gradually emerged with the modern engineering profession. Engineering[12] developed first as a part of military activities and later became separate from the army association. Among the first engineering education institutions were the Moscow School of Military Engineers (established 1698), the Apprenticeship School for Civil and Military Engineers in Prague (1707), and *Bureau des dessinateurs du roi* in Paris (1744). In Britain, engineering education took no independent institutional forms but was installed in existing colleges like King's College in London (1838). The United States Military Academy at West Point graduated the first engineers in 1817. The first private civil engineering college in the United States, Rensselaer Polytechnic Institute, was founded in 1823.[13]

In the United States, engineering ethics has assumed the role of creating identity, stature, and prestige for a profession that lacks the long traditions of clergy, medicine, and jurisprudence. In Europe, the discipline has also reached for broader perspectives, attempting to ask general questions concerning technology and the good of society. This is a rough characterization, however, since engineering cultures, institutional arrangements, and ethical thinking within engineering take various forms in European countries and globally.[14] Issuing codes of ethics for engineers has been an important way to build up the character of the profession. These codes were, and still are, drafted and announced by various engineering societies. In England the first codes were issued in the middle of the 19th century, and in the United States, in the beginning of the 20th century. In continental Europe the codes were made official only at the turn of the millennium.[15]

The content of engineering ethics codes varies, but most codes currently focus on the obligations to public safety, health, and well-being. Yet this has not always been the case: before the 1970s the primary obligation was to the employer or client.[16] The tension between these obligations, to the employer or to the public, becomes concrete in the issue of whistle-blowing: When is an engineer obliged to reveal to the public a wrongdoing (or a problematic policy) at his or her institution?

Two cardinal examples of whistle-blowing are the C-5A cargo plane and the DC-10 passenger plane incidents. In the C-5A case, engineer Ernest Fitzgerald told the truth about the cost overruns in Senator William Proxmire's hearing. In thus doing, he acted against his superiors' will and the interest of Lockheed Aircraft Corporation. Honesty was costly for him, as his career was destroyed. The tragic episode lasted from 1968—when the whistle-blowing took place—until 1981—when Fitzgerald finally got his job back. In the DC-10 case, whistle-blowing, had it been successful, might have saved lives. In 1974 a DC-10 crashed over the Paris suburbs, and 364 people died. The weaknesses that caused the crash were known before the accident. Senior engineer Dan Applegate had reported two years earlier to the vice president of Convair—a subcontractor of McDonnell Douglas—the safety problems in the DC-10's cargo doors and cabin floor. He recommended redesigning the doors and strengthening the floor. The company officials agreed on the technical details but decided that for financial and contractual reasons, the improvements were impossible to realize.[17]

The notion of whistle-blowing brings a tangible moral dimension to the engineering praxis, but at some cost. The whistle-blowing narratives create an individualistic picture of a moral hero who sacrifices her or his personal interests for the right cause, a picture that may give an inadequate account of the social, institutional, and corporate character of engineering praxis. Furthermore, depiction of the responsible conduct as supererogatory moral heroism or martyrdom hardly makes it palatable to engineering students.[18]

In addition to whistle-blowing, issues in engineering ethics include imposition of risks, conflicts of interest, confidentiality, environmental effects, and tensions between different cultural practices and traditions.[19] A classic case of decision-making in engineering was the *Challenger* catastrophe. The engineers of Morton Thiokol, the company that produced the booster rockets for the shuttles, suspected that the weather on January 28, 1986, was too cold for the O-ring seals of the rockets and, therefore, launching the shuttle was risky. The managers of the company disagreed and, in the end, persuaded the research engineers to accept the decision to launch. The O-rings failed, and the shuttle exploded 76 seconds after launching.[20]

Academic research, teaching, and writing on engineering ethics have increased since the early 1980s. The methodology of engineering ethics has followed the general developments in applied ethics. The three basic approaches have thus been a theory-based "top-down" model, a "bottom-up" model that takes off from concrete situations, and a "mid-level" model that anchors ethics in principles, for instance, the principle of nonmaleficence, without seeking an overall theoretical backing for them.[21] Some add to these two context-related models: feminist ethics and virtue ethics. Although such approaches to engineering ethics are possible and relevant, they have rarely been advocated.

Critics have, on the one hand, found engineering ethics too individualistic and neglectful of a public-good perspective.[22] On the other hand, some have found it awkward that engineers should think about their profession and its goods very widely, as society does not expect such reflection from most other professions: judges, for instance, do not write the laws, and military professionals do not decide to go to war.[23]

As noted in passing, engineering and technology initiate numerous and large-scale transformations of the environment. In fact, humans would have been unable to end up producing most environmental disasters without modern technology. Environmental ethics, therefore, addresses questions closely related to technology. The academic discussions on environmental ethics began in the early 1970s, and the field achieved considerable growth in the '80s and '90s.[24] Even before this formative period some serious ecological considerations had been raised. Ralph Waldo Emerson (1803–1882), Henry David Thoreau (1817–1862), John Muir (1838–1914), Albert Schweitzer (1875–1965), and Aldo Leopold (1887–1948), for instance, put forth propositions for a new, more ecologically benign kind of thinking.[25]

The issues of environmental ethics are numerous. Yet some central problems are present in most discussions. The main problem is the concrete and conceptual tension between humanity and nature. Environmental and ecological damage clearly result from the human domination of nature. This domination has its conceptual ground in views that separate human beings from the rest of nature and perceive the natural world as a usable resource for humanity. Remedies for the ecological predicament, therefore, appear to require radical rethinking of the relationship between humankind and nature.[26] Although human-centered approaches to environmental ethics exist, this line of argument often entails questioning the anthropocentric worldview that has been dominant in our culture. Sentientism (the sphere of sentient beings), biocentrism (life in general), and ecocentrism (the ecosystem in its totality, animate as well as inanimate) are the possible alternatives for human-centeredness. Regarding ethics, however, extension outside the human realm is in many ways problematic. Usually, morality is thought to exist between moral agents who can be expected to act morally toward each other and to share duties and rights reciprocally. Even sentient beings are not able to form a moral relationship to humans or amongst themselves. How reasonable is it, then, to think that they have rights that entail human duties? Their ability to feel pain could be a basis for these nonreciprocal rights. The problem, nevertheless, extends further: Sentient beings are a small minority in the whole of life, and the inanimate parts of the ecosystems face destruction, as well. If nature has only instrumental use-value for humans, it appears that few limits for harsh exploitation of ecosystem and destruction of life exist. Should environmental ethics then find some intrinsic value in nature? The suggestion seems tempting but raises problems, as well, since if humans seem to be the only value-positing beings, the value of nature depends on the subjective human value judgment. This would be both an anthropocentric and arbitrary basis for the intrinsic value in nature.[27]

The above reflections reveal a glimpse of the troubles related to thinking about nature and morality in conjunction. Although the discussions in environmental ethics often reach out for big questions, they still need to infuse more of the world-revealing and world-making powers of science and technology into the picture.

After the Second World War, medical-technological innovations such as kidney dialysis, organ transplantation, safe abortion, contraceptive pill, respirator, prenatal diagnostics, and genetic engineering raised novel social and moral questions. They concerned boundaries of birth, life, personhood, and death. Moreover, new medical technologies were expensive, which forced institutions to reflect on the just distribution of scarce resources. As academic philosophers and theologians in cooperation with physicians began to address these issues, modern biomedical ethics originated. Obviously, physicians have for centuries possessed an explicit moral code in the form of the Hippocratic Oath. Conscious academic reflection on concepts, practices, and goods in the medical profession, however, has a history starting only in the 1960s. In Anglo-American philosophy the emergence of biomedical ethics constituted the first major wave of reviving interest in practical moral questions. Previously, academic philosophy had mainly concentrated on abstract analysis of moral language, or metaethics. In the 1970s and '80s bioethics took considerable formative steps as an academic field. Currently, bioethics classes are taught in all major research universities in the United States, and the field is developing in all advanced societies.[28]

The notion of "bioethics" is considerably unclear and under dispute. In the narrowest sense, bioethics equals professional ethics in medicine. The broadest understanding of bioethics perceives it as ethics concerning all life, especially in those respects that have become manageable through the life sciences and biotechnology. This latter understanding converges conceptually with environmental ethics. In contingent academic practice, however, these two disciplines have remained separate. The most common conceptions of bioethics exist between the two extremes.[29]

From Ethics to Politics

While various applied ethics fields within their scope successfully address moral issues, wider and more political questions in relation to technology exist. Next, I will briefly review some of the most important. These questions could obviously be included in the scope of bioethics or environmental ethics when they are broadly cast.

First of all, how are the benefits and costs of any technology distributed among individuals, countries, and globally? This is a question of social justice. In connection to modern technology, these issues have a history that begins with the exploitation of the working class in the early phases of industrial revolution. Although few people any longer apply the Marxist theories of class struggle to the past or the present, it appears obvious to most that the reforms pleaded by the labor movement in the late 19th and early 20th centuries were desperately needed. In the current developed societies, the dismantling of welfare policies has launched new discussions on the just distribution of technologically produced goods.[30] Justice on a global scale is yet another pressing question. Wealth and access to resources are very unevenly shared among the peoples on the globe. Economic globalization, moreover, forces nations and other actors to reflect on just distribution in rapidly altering conditions.[31] Economic agents may make use of disparities in the costs of raw materials, labor force, taxes, laws, and regulations. At the moment, an academic and political debate concerning justice in globalization goes on.[32] Whatever the outcome of the debate, child labor, human trade, and waste dumping in developing countries, at least, are serious moral problems.

Cost-benefit analysis is the most common framework for socially assessing technological projects. "Costs and benefits" refer to the known and intended consequences of technological projects. These most often, however, have some unintended consequences as well. For instance, the intended consequence of DDT was to work as a potent insecticide; the unintended consequence was its effect on other animals and the accumulation in the food chain.[33] The unintended consequences that can be framed and assessed are called risks. A risk is a function that combines the probability and magnitude[34] of an adverse effect. Yet one can easily perceive that room for ambiguity remains between the model and reality. Risk assessment, in fact, raises a number of epistemological, ethical, and political questions concerning the objectivity of risk, emphasis, and value given to different factors and the due process of risk assessment. Moreover, all the unintended consequences can seldom be taken into account. Therefore, uncertainty remains a relevant and realistic notion in the context of technology.[35]

Technological vulnerability is the complementary concept for technological risk. "Vulnerability" refers to the brittle character of technological systems. Nuclear power plants and microcircuits, for instance, are vulnerable to outside influences, and these systems going into jeopardy creates a risk for society. After September 11, 2001, the concept of vulnerability gained a new meaning.[36]

A more comprehensive method than risk assessment for political decisions concerning technology is technology assessment (TA). Kristin Shrader-Frechette characterizes it as follows: "A key component of TA is the evaluation of social, ethical, and environmental impacts likely to be associated with a technology; an environmental impact assessment is one part of a TA."[37] Yet achieving the task of TA is difficult: it calls for successfully blending scientific and political, objective and subjective "facts" and "values," as well as humanities and natural sciences. The most well-known TA offices include the U.S. Congressional Office of Technology Assessment (OTA),[38] the Rathenau Institute in the Netherlands, the Danish Board of Technology (*Teknologirådet*), and the Japan Science and Technology Agency.

Shrader-Frechette describes the content and procedure of technology assessment as follows:

> TA typically includes at least ten components: (1) a problem definition; (2) a technology description; (3) a technology forecast; (4) a social description; (5) a social forecast; (6) an impact identification; (7) an impact analysis, including risk analysis; (8) an impact evaluation, including risk evaluation; (9) a policy analysis; (10) communication of results.[39]

However, both technologies and social contexts vary to the extent that no uniform TA methodology for accomplishing these stages exists. Actual technology assessments have been criticized for different shortcomings. Neglect of both implicit and explicit ethical dimensions in assessment; underestimation of cultural, political, and social impacts; failure to produce unbiased results; and the lack of a comprehensive approach—especially due to certain quantitative methods such as probabilistic risk assessment (PRA) and cost-benefit analysis (CBA)—have been raised as weaknesses of TA. Despite these problems, technology assessment provides a forum for the systematic evaluation of technologies prior to their implementation. The relevance of ethics in TA follows from the need to analyze TA's central moral concepts, such as consent, freedom, and equity. Further, identifying implicit moral frameworks—utilitarianism vs. deontological approaches, or libertarianism vs. communitarianism—in TA is a task for the ethicist. Finally, because of the kind of forum it is, TA presents a locus for evaluating how ethical certain use of technology, in fact, is.[40]

Despite the conceptual issues in technology assessment and risk analysis, decisions concerning technologies are made all the time. In this context, it is a question of participatory justice who is included in the decision-making process and how. On the one hand, a common view would bestow the decision-making power on those who invest in the projects and on their experts. On the other hand, some have invoked "free and informed consent," a notion from bioethics, to advocate public participation. Free and informed consent attempts to capture the conditions under which a patient or human research subject can be thought to give an autonomous assent to a measure with relevant understanding.[41] Some have argued that issues of informed consent are not limited to human subjects of biomedical research but occur in the citing of chemical plants, the construction of major transportation facilities such as airports or highways, and consumption of technologically modified foods.[42] This kind of demand sounds reasonable. It is, however, far from clear in what kind of public sphere the due process for acquiring the informed consent could be organized.[43]

Attempts at creating appropriate spheres for decisions on technology were made around the turn of the millennium. Constructive technology assessment (CTA), for instance, presents a further development of the technology assessment methods in a direction that admits involving a number of different interest groups. These groups can include consumers, citizens, corporations, employees, local inhabitants, and other social groups. CTA makes use of consensus conferences, dialogue workshops, citizens panels, and scenario workshops, for example, to bring different actors together.[44]

CTA possesses the obvious advantages of increasing transparency, public participation, and social learning. It does not, however, present a magic solution for the question of how to think about ethics in decisions of technology: delegating questions to "the public" is not a solution. Hypothetically, the citizens, as participants of a consensus conference, for instance, may call for assistance from the people and institutions that study ethics. In short, the decision procedures should be able to balance the requirements of democracy/participation, various kinds of expertise, and consistent ethical/political thinking.

All of the above reflections rest on an assumption that at least some actors have autonomy in making decisions concerning technology. For instance, the request for an increase in citizens' participation assumes that some actors—corporations, bureaucrats, or experts—are making decisions in which the larger populace should have a say. No consensus, however, exists among social theorists concerning to what extent and by whom technological development can be controlled or steered. The notorious notions of technological determinism and the autonomy of technology claim that technoscientific development has its own laws of motion, in the workings of which humans have little or no power. For obvious reasons—the notions evade the demands of falsification—these views lack adherence. The concrete examples against technological determinism, instances of an actual human steering of technological development, are, however, not so easy to point out.[45]

Even if the ideas of technological determinism and the autonomy of technology are proven unfeasible, in a more limited sense the ethical-political character of technological artifacts still remains to be reflected upon. The artifacts act as parts of our social contexts. The clearest examples of this are various automated technological devices and systems such as robotic construction lines, autopilots in airplanes, and computerized warfare systems. More ordinary examples include various electronic platforms for job applications, banking, dating, and social life in general. Although these systems do not deliberate morally on their doings, they set boundaries for human agency. These boundaries obviously are most often inscribed by human designers and programmers—within the possibilities of hardware and software. Yet the automated devices stretch the human proxy far and act quite independently based on it.

In a slightly different vein, scholars have discussed in what sense artifacts may have political characteristics.[46] These characteristics may exist in at least two distinct manners. First, technological artifacts may be externally used as a medium for advancing political interests. The design of public transport systems, for instance, structures the mobility of different populations. Second, some technologies may be inherently bent to favor certain political arrangements. Nuclear power, for instance, with its close relationship to nuclear weapons as well as dangerous fuel and waste management, makes secretive and forceful measures necessary. These measures are in tension with democratic ideals.[47]

Although the questions concerning the technological artifacts' ability to be active on their own and their possible political characteristics are related, they are clearly separate, as well. It is especially worth noticing that artifacts need not be able to "act" in order to be politically charged. Most examples of artifacts having political features, in fact, stress more the structuring power of technologies than their ability to perform autonomously.[48]

In case human autonomy in steering technological development exists, at least the following ethical-political questions arise: Who controls the technological change? Who has the right to participate in steering the development? How should different stakeholders in decision-making procedures concerning technologies be proportioned? These stakeholders include at least those who possess the capital to invest, legislative bodies, policy makers, professional experts, and those citizens whose lives are affected by the application of technologies. These questions exist on the margins of a political discourse that mostly emphasizes more traditional power relations instead of looking at the moral and political significance of the material contexts of communal life. The steering of technological development, nevertheless, encompasses a genuine task for political thinking: how to reconcile the tensions between economic competition, due democratic processes in decision-making, and narrow, intense (technical, managerial, or bureaucratic) expert knowledge?[49]

To a slightly more general level ascends the question concerning technology and the good life. Today, the question is seldom explicitly asked. The reason for this absence of questioning appears to be twofold: either it is thought that no such question exists—the good life belongs to the sphere of moral questions, and technology does not really relate to morality—or it is taken for granted that better technology means better life. In the past, however, this question has been more open. In the beginning of the modern era, a new perspective emerged in science and philosophy. Thinkers who participated in the formation of the novel natural scientific framework explicitly held that this pragmatically oriented technoscience would provide humanity with a better life. In *Discourse on Method* (1637), René Descartes articulates this ethos as follows:

> Through this philosophy we could know the power and action of fire, water, air, the stars, the heavens and all the other bodies in our environment as distinctly as we know the various crafts of our artisans; and we could use this knowledge—as artisans use theirs—for all the purposes for which it is appropriate, and thus make ourselves, as it were, the lords and masters of nature. This is desirable not only for the invention of innumerable devices which would facilitate our enjoyment of the fruits of the earth and all the goods we find there, but also, and most importantly, for the maintenance of health, which undoubtedly is the chief good and the foundation of all other goods in this life.[50]

Moreover, during the Enlightenment thinkers linked technoscientific progress to political progress, manifest in expansion of liberty, equality, and democracy.[51]

The promise Descartes and others presented has largely come to pass: technoscientific progress has enabled the wide-scale spread of democracy, well-being, and equality—in developed societies. Moreover, this same advance appears to have granted us greater freedom in terms of self-realization.[52] This development, however, has created serious problems, as well. The ecological crisis that has followed technological mass production is, arguably, the most severe manifestation of this predicament.[53] Although at first glance the promise of technology, with respect to human life, appears to have proven true and beneficial, upon closer examination the situation is more ambivalent. Since the late 19th century many cultural critics have expressed doubts concerning the nobility, integrity, and wisdom of the way of life in modern societies.[54]

Ample but fragmentary empirical evidence in support of these doubts exists. For instance, although institutional science progresses, it is not evident that people's scientific worldview is in fact becoming more accurate. Moreover, it is not clear whether citizens in developed societies profess to be happier than people in underdeveloped countries. The public awareness of communal, national, and global issues does not appear to improve much despite the apparent technologically created possibilities for being informed.[55]

Reading the corpus of social scientific research data concerning these trends is an interpretative task, since the researchers most often do not evaluate the state of the society as a whole but rather confine themselves to describing small regions of social life.

One can make certain divisions in the nature of the challenges that technological development has posed to ethics. A useful classification divides them into personal, professional, and public spheres.[56] Engineering ethics foremostly addresses professional moral questions common to engineers. The way in which engineering ethics relates to the public good is disputable: How much should the engineering profession reflect on the goods it provides? The clearly public moral questions concerning technology include legislation, public funding of technology projects, equality and inequality in the context of technology, and general technology policy. Personal questions concern individual understandings and uses of technology. At first sight, this category seems to consist of consumer choices and personal convictions regarding the good life. A closer look at these matters, however, reveals that public and personal intertwine: delegating energy decisions, for instance, to consumers may be a political move. In fact, the development that subdues citizens under the notion of consumer remakes the boundaries of what is public and personal.

APPLICABILITY OF ETHICAL CONCEPTIONS

In the previous section the focus was on phenomena that may be framed as challenges for ethics posed by technoscientific advance. Looking at these challenges, one can perceive that adequate treatment of them forces attention to the ethical concepts and frameworks, as well. Technological mediation alters the character of human action and moral agency, as the example of killing and military technology demonstrates. Thus a question arises: How adequate as tools are the traditional Western ethical concepts and theories in the late-20th-century and early-21st-century world? Contractarian, deontological, and utilitarian theories in ethics do carry the name of modern ethical theories. Yet they have little to say about the human condition in technological contexts and about artifacts mediating human action. Next, I will introduce some issues concerning the applicability of traditional ethical concepts in a technological world. After that, I will proceed to a broader framework that classifies human action in respect to its moral relevance.

A first set of issues relates to the way central ethical concepts are understood. For instance, the concept of harm is most central in ethics, since even a most basic ethical code consists of the principles of beneficence—to promote good—and nonmaleficence—to avoid harm. But what is harm? In premodern times harm was conceived rather unproblematically as a tangible and spatiotemporally immediate damage. In contemporary society this is not quite so simple. Can the noise of the jet planes above my suburb be counted as harm to me? How can long-term exposure to pollutants with consequences that are framed within probabilities be considered harm?[57]

The concept of harm and its meaningful application in contemporary society, in fact, represent an overall task for ethics: in many ways ethical thinking needs to be in dialogue with scientific knowledge. This is especially true in public policy and institutional ethics but holds in personal ethics as well. Such a need for scientific knowledge was nearly absent in the ethical traditions of the pretechnological world. As modern a thinker as Kant makes a clear remark on the redundancy of knowledge in ethics: "thereby do we show that neither science nor philosophy is needed in order to know what one must do to be honest and good, and even wise and virtuous."[58]

Furthermore, the concept of right is essential in the traditions of liberal societies. It appears natural to think that no one's rights should be limited unless they violate the similar freedom of the others. The powers and abilities of agents—individual and institutional—have, however, been radically enhanced due to technological development. Does the traditional concept of right apply in these conditions? For instance, what happens if there is no limit to our right to purchase and use powerful polluting devices, such as big private cars? How can the right of corporations to do business be reconciled with environmental concerns?[59]

In addition to critically reflecting on traditional concepts, another way to regain adequacy in ethical thinking is to call attention to new concepts. One such concept is 'responsibility.' As a 300-year-old modern concept, responsibility made its way to the center of the discourse on science, technology, and ethics during the 20th century. The word derives from Latin *respondere*, "to promise in return," "to answer." The English verb "to respond" and the adjective "responsible" preceded the abstract noun. The pretechnological Western culture had a locus for such a concept: in Judeo-Christian-Islamic tradition the response of a human being to a call from God has a central religious and moral importance, which is also enacted in the "responses" of liturgical practice. Despite this evident locus, responsibility became a part of religious discourse only through Karl Barth's (1886–1968) work. After Barth invoked it, responsibility became central in both Protestant and Catholic theology in the 20th century.[60]

In addition to religious meanings, the concept of responsibility is applied in legal discourse. There, "liability" equals responsibility. Both in criminal and in civil law, an essential distinction concerns the intentionality of a transgression. Liability is considered to be lesser in unintentional acts than in intentional ones: an accidental homicide is less severely punished than a deliberate murder. In the contexts of technological undertakings, however, the boundaries of liability have been slightly transformed, and manufacturers are also considered liable in tort when their products have unintentional defects that cause damage or injury.[61]

Philosophers have thought about responsibility from various angles. One perspective emphasizes that such a notion was needed after the breakdown of hierarchical social and metaphysical orders that had served as the bases of duties. A second theme is that the causal and temporal frames of modern technology required a new kind of principle to capture human duties in the context of increased human powers. The single technology that raised the most discussions on the responsibilities of scientists and technologists was the atomic bomb and other nuclear weapons that followed it.[62]

As we noticed in the case of engineering ethics, although the responsibilities of engineers have been under dispute, the responsibility for securing public safety, health, and welfare begins to be widely accepted. The responsibility in engineering is more concrete than in science, since engineers aim at explicit goals as they create tangible products, while the results of science are less predictable.

Responsibility is closely associated with the issue of control touched upon previously. Since the modern technological networks have a vast multi-actor—including both numerous human actors and technical devices—character, attribution of responsibility within these networks is seldom simple. Furthermore, contemporary technologies involve time frames and causal connections of such a novel scale, the responsibility that comes with using them must be conceived quite differently from the earlier notions of accountability that had to do with more immediate and tangible relationships.[63]

The broad dimensions of responsibility in technological action relate to thinking about the recipients of the action. For the agents, the recipients are often to some extent unknown or uncertain: in vast causal-temporal frames they are nearly impossible to identify fully. They may even not yet exist. The biggest discussion, in fact, revolves around the issue of harming the living conditions of future generations. The present generations may destroy the environment so that it threatens the life of future generations. Cross-generational responsibility, however, is a novel concept in ethics and politics, as the existence of the recipients of the action is contingent. Moreover, the size of the next generation is in novel ways dependent on actions of the present generation, due to modern medicine and contraceptives. In some sense the present generation is responsible for both the size of the next generation, through population policies, and the ecological predicament that is left for it to survive with. The grounds for the moral obligations in cross-generational relationships and the theoretical frameworks for identifying them, not to mention the detailed analysis of the issues, are so far works in progress without a consensus.[64]

As we noted previously, the agent as well as the recipient of technological action raises questions. No clear line divides human beings from technological devices and systems in questions of moral and political agency. Technological artifacts can both act by themselves in varied independence and structurally shape human agency. The category of cyborg,[65] a human-machine mixture, becomes more and more real in ordinary life, apart from science fiction: contraceptives, pacemakers, cellular phones, personal music devices, implants, prostheses, and functional foods present cases in which clear distinctions are blurred.[66] Even transportation devices such as bicycles and cars alter the agent's relationship to her surroundings and create a new actor: a human-device combination in which the human part makes the choices but the device part sets limits for choosing the effects and their magnitude. An anti-locking braking system in a car, for instance, changes the conditions for the human pursuit of safe driving.[67]

This questioning can go deeper, when we consider the ways in which humans can be technologically molded. Plastic surgery is a superficial form of such molding. Medicines of the psyche represent softer and more delicate technological fixes that alter the human being. Xenotransplantation, moreover, has forced us to think about the boundaries of species. Gene modification may alter these boundaries radically in the future. In the context of bioengineering, the merging of human and technological often means a fusion not of "a man and a machine" but of human organism and artificial organic creations.

All in all, these observations show that questions concerning human nature are at the core of the discourse on ethics and technology. To begin, there are notional questions about how the human being is conceived—materialistically or dualistically, as *homo sapiens* or as *homo faber*, for instance.[68] These ideal conceptions have practical bearings on how human beings are treated in education, political life, and employment. Some social critics argue that the technological era entails a specific and "dehumanizing" understanding of human beings.[69]

Additionally, there are the various material alterations of human beings and their relation to their habitat. This openness of anthropological questions has urged people in the academy to speak about transhumanism[70] and the possibility of a posthuman future.[71] "Transhumanism" refers to the idea that humankind should deliberately with technological means transcend humanity as it is now known. Transhumanists advocate overcoming aging, sickness, disability, and suffering with available technologies. They often also think that humankind should without inhibitions enhance its natural abilities, for example, intelligence and memory. "Posthuman" is a term more often used by the critics of transhumanism to denote the condition after traditional humanity. It, moreover, appears that the posthuman condition might inevitably result from "the normal development" in the liberal Western societies, as possibilities for regulating technologies are few.

In this territory vast speculative views open. Although no metaphysical consensus on human nature is likely to be achieved, ethics that address technological phenomena must pay attention to the ideas of *human well-being* that underlie policies and development.[72]

Various individual notions point at a broader problem in applying traditional ethics in technological culture. The problem has to do with the distinction between production and moral action. Since Aristotle, human activities have usually been divided into three types. The first type, productive actions, making, or *poiesis*, is considered morally neutral. The second type, doing, or *praxis*, is perceived as the medium of moral questions. The third type, theoretical and scientific action, or *theoria*, is again outside moral concerns.[73]

THE MEANING AND FOUNDATIONS OF ETHICS

In the modern situation these distinctions appear less and less viable. If the observations I have made concerning ethical issues connected to technology—the questions of distribution, risk, good life, agency, and so on—are

Table 2.1. Levels of Revision in Ethical Notions

Levels	Examples	Issues	Possible Revisions
Concept	harm	causal and temporal frames of a novel scale	precautionary principle
	right	increased human powers and broad rights, a problematic combination	responsibility, a modern notion
Theory	virtue ethics contract theories deontological ethics utilitarianism	increased human powers and technological mediation alter human *agency* moral decisions require knowledge in a novel manner	taking formation of agency into account in ethics
Frame-work	*poiesis-praxis-theoria* distinction humans as the only possessors of moral choice anthropocentris m	in concrete reality, technological devices and scientific practice interpenetrate production of technological devices appears to have moral implications technological devices function as active parts of organizations and systems animals suffer ecosystems lose their balance	creating ways to assess things morally different forms of technology assessment (TA) thinking about moral characteristics of technological products; attempts to take agency of artifacts into account extending moral responsibility to animals considering intrinsic value in nature

correct, then we cannot preserve the distinction of moral *praxis* and amoral *poiesis*. In other words, a crucial obstacle for thinking about ethics in technology is the notion that manufacturing is in itself morally neutral.[74] If this conclusion is correct, some means should be found to morally evaluate technological development in all its phases, from planning and designing to finished products and structures.

Not only is the morally neutral concept of poiesis problematic, but the value-neutral notion of *theoria* is actually as questionable. In fact, often in political discourse, technology is legitimated on the grounds of its being based on science. Since science is an objective description of reality, the argument goes, and technology is based on scientific research, there cannot be many value questions concerning the design and production of technologies. Thus, ethics and politics come into the picture only after production, downstream, and mainly address the questions of how the technological devices are used.[75]

Social studies on science and technology show that this reasoning is simplistic: science and technology cannot be cleanly separated. Furthermore, research and development are perceived as increasingly important forces of production. Due to this, the technoscientific complex is both a receiver of strategically and quantitatively significant investments and a producer of equally important output. Certainly, then, where the investment is allocated and what kind of production is targeted must be questions of valuation. This kind of critique of value-free science is valid irrespective of the general philosophy of science. Even if one professes some kind of realistic view of science, one must explain how science as human action and as resource allocation could be value free.[76]

Table 2.1 summarizes the need for revising ethical notions on different levels. In brief, the questions of adequacy and applicability have to do with accommodating ethical thinking in conditions altered by technological advance. Reflection on accommodation usually works on the level of detailed concepts, central theories, or frameworks. No serious attempts to create a whole new ethical theory for the technological age have been made.[77] Instead, more attention has been paid to allegations that have deemed ethics obsolete in the era of science and technology. These arguments are the topic of the next section.

The previous two sections looked from different perspectives at the possibilities of applying ethical thinking to the moral questions emerging in a technological world. In the past it was mostly taken for granted that such ethical thinking was meaningful and relevant, not trivial or arbitrary. Modern technoscientific development, however, has in certain ways questioned the relevance and foundations of ethics. Karl-Otto Apel writes:

> The same *rational, scientific* thinking that makes technology possible and so constitutes the external provocation for an ethic of responsibility,—that same thinking as a *value-free objectification of the world*, seems to show a *rational founding of ethical norms* to be impossible.[78]

Philosophers and ethicists have reported on this questioning at times triumphantly, at times lamentably. Below, I sketch the situation of ethics in the main philosophical traditions of the 20th century: analytic philosophy and phenomenological tradition. This sketching emphasizes the relationship between technoscientific development and ethics; it centers around the towering figures of those traditions: Ludwig Wittgenstein (1889–1951) and Martin Heidegger (1889–1976).

Problems of Ethics in Analytic Tradition

The bluntest articulation of the irrelevance of ethics in the technoscientific culture came in the 1920s and '30s from logical positivism. This movement aimed at creating scientific philosophy and thus came to present an intellectual articulation of the achieved state of disenchantment.[79] The criteria, namely distinct empirical content and transparent logical structure, that logical positivism set for the meaningful use of language ruled ethics—with metaphysics, religion, and aesthetics—out of the category of statements with a real semantic content.[80] A. J. Ayer describes his view of ethics as follows:

> We shall set ourselves to show that in so far as statements of value are significant, they are ordinary 'scientific' statements; and in so far as they are not scientific, they are not in the literal sense significant, but are simply expressions of emotion which can be neither true nor false.[81]

Accordingly, a consequence of this demarcation was to regard ethical sentences as expressions of the speaker's emotions, i.e., metaethical emotivism. According to emotivism, for instance, the meaning of the sentence "murder is wrong" would be something like "I disapprove of murder" or "I don't like murder."[82] Ayer analyzes moral sentences as follows:

> Thus if I say to someone, 'You acted wrongly in stealing that money,' I am not stating anything more than if I had simply said, 'You stole that money.' In adding that this action is wrong I am not making any further statement about it. . . . For in saying that a certain type of action is right or wrong, I am not making any factual statement, not even a statement about my own state of mind. I am merely expressing certain moral sentiments. . . . It is worth mentioning that ethical terms do not serve only to express feeling. They are calculated also to arouse feeling, and so to stimulate action.[83]

Although this extreme stance toward ethics softened gradually within the logical positivism that formed the core of what developed into analytic tradition in modern philosophy, the reservations concerning normative ethics remained. Hence, analytic philosophers regarded metaethics, which analyzes the logical properties of moral parlance without a contentual commitment, as philosophically a more relevant pursuit than ethics in a traditional and comprehensive sense.[84]

Ludwig Wittgenstein worked in close contact with logical positivism, yet his thinking was in many respects independent from the movement. On the one hand, in *Tractatus Logico-Philosophicus* (1921), he clearly stated that ethics falls out of the sphere of scientific and philosophical usage of language. On the other hand, Wittgenstein considered ethics extremely important. This became evident both in his texts and in personal correspondences and conversations. In fact, exactly because ethics has to do with matters absolute, the words of the finite language are unable to capture the demands of ethics. In Wittgenstein's thinking the ethical comes close to the mystical. This is how he characterizes ethics in *Tractatus*:

> 6.4. All propositions are of equal value.
>
> 6.4.1. The sense of the world must lie outside the world. In the world everything is as it is and happens as it does happen. In it there is no value—and if there were, it would be of no value. If there is a value which is of value, it must lie outside all happening and being-so. For all happening and being-so is accidental. What makes it non-accidental cannot lie in the world, for otherwise this would again be accidental. It must lie outside the world.
>
> 6.4.2. Hence also there can be no ethical propositions. Propositions cannot express anything higher.
>
> 6.4.2.1. It is clear that ethics cannot be expressed. Ethics are transcendental. [Ethics and æsthetics are one.][85]

Wittgenstein's only text that primarily addresses ethics is "A Lecture on Ethics," which he delivered at the end of 1929 or the beginning of 1930 to the Heretics Society. This lecture concludes as follows:

> My whole tendency [in "A Lecture on Ethics"] and I believe the tendency of all men who ever tried to write or talk Ethics or Religion was to run against the boundaries of language. This running against the walls of our cage is perfectly, absolutely hopeless. Ethics so far as it springs from the desire to say something about the ultimate meaning of life, the absolute good, the absolute valuable, can be no science. What it says does not add to our knowledge in any sense. But it is a document of a tendency in the human mind which I personally cannot help respecting deeply and I would not for my life ridicule it.[86]

In his later philosophy Wittgenstein receded from many starting points of *Tractatus*. Most importantly, he began to think about the semantics in a much richer manner than he had earlier. He developed a way to understand verbal meaning that emphasized the use of words in the contexts of life, pragmatics more than syntax or formal semantics. In *Tractatus* Wittgenstein had created a picture theory of language, according to which sentences of a language represent the world in a quite literal, picture-like manner.[87] In *Philosophical Investigations* (published posthumously, 1953) the meaning of a word or sentence follows from its place and role in the language game (*Sprachenspiel*).[88] In short, this way to understand meaning emphasizes both saying as action and the contextual relatedness of meanings, instead of the earlier, more static, atomistic, and representational view. A form of life (*Lebensform*) sets the transcendental conditions for a language game.[89] For instance, the verb "to demonstrate" has a certain meaning within the natural scientific language game. This meaning relates to other words used in that game, and demonstrating is also something that one does within the language game. The game rests on, or within, the form of life called science, which consists of practices the scientists do, things taken for granted, the way to see the world in "the light of science." One cannot determine an exact semantic meaning for the verb "to demonstrate" without taking into detailed account these practical and contextual perspectives.

At first glance, Wittgenstein in *Investigations* appears to make ethics relative to forms of life and language games. It seems that different forms of life are incommensurable. There might well be a language game of morals, but it would be confined to its own sphere. Judgments across different language games would be problematic. A religious person, for example, perceives a scientist's critique of her beliefs as "not demonstrated to be true" misplaced, since religious convictions are usually not demonstrated. Wittgensteinian analysis of the situation would say that in this case, the scientist is forcing the scientific language game beyond its proper sphere. Wittgenstein's later philosophy, in fact, has had among its interpretations this kind of relativistic trajectory. Wittgenstein himself, however, remained silent concerning ethics after "A Lecture on Ethics."[90]

* * *

Wittgenstein studied mechanical and aeronautical engineering before getting engaged in logic and philosophy. He also practiced architecture during his active years in philosophy. He, however, never directly addressed the problems of technology in any works meant to be published. In his notebooks, he nevertheless expressed deep worries about "the technoscientific civilization," for example, in the following manner:

> The truly apocalyptic view of the world is that things do not repeat themselves. It isn't absurd, e.g., to believe that the age of science and technology is the beginning of the end for humanity; that the idea of great progress is a delusion, along with the idea that the truth will ultimately be known; that there is nothing good or desirable about scientific knowledge and that mankind, in seeking it, is falling into a trap. It is by no means obvious that this is not how things are.[91]

Despite Wittgenstein's (public) silence on technology, one could well use his central concept to characterize life in the contemporary world: one could say that we live within a technological form of life.[92] In the same way that religion characterized the form of life in the Middle Ages, now that Gothic cathedrals have given way to space shuttles as landmark projects, it appears that contemporary culture is characteristically technological. Since the form of life is the basis of the language games, living in a technological form of life results in semantics becoming technological: the meanings of words tend to turn technological. Consider words such as "input," "output," "feedback," and "resource" used in human contexts. In a psychotherapeutic doctrine called neurolinguistic programming (NLP), human psyche is understood as programmable system.[93]

Analytic tradition remained rather satisfied with its mostly metaethical pursuits until the last third of the 20th century.[94] Then, however, the need for ethical expertise in medical practice became evident. The moral questions posed by biomedical dilemmas uncovered the shortcomings of abstract meta-ethics. As philosophers and theologians were called upon to consult with physicians in medical praxis and in new situations that emerged due to medical technology, they were rather unequipped for these practical tasks. Academic philosophy was an inadequate tool for tackling concrete moral dilemmas. Soon other areas of human endeavor followed medicine in assigning tasks for ethicists. Since then, applied ethics has become a major field in academic philosophy and theology.[95]

The field of applied ethics, nonetheless, still lacks credibility and rigor.[96] Symptoms of this condition include at least the following: First, the theoretical foundations of applied ethics are unclear. Often this lack of theoretical backing is embraced with favor: casuistic situation ethics claims that moral thinking is best construed bottom-up, not top-down, from general theories, and "principlism" trusts the mid-level principles, not grounded in any larger theory, as the core of applied ethics.[97] Obviously, forms of consequentialism and deontological ethics are still available for general theoretical frameworks—they have, in fact, seen a revival since the late 20th century. Yet applied ethics appears to continue with the unclear theoretical profile that marked its beginnings. Second, the issue-centered perspective of applied ethics appears politically naïve: applied ethicists most often are not asking questions about policies and broader politics that become concrete in the issues they study.

Finally, some Wittgensteinian philosophers have leveled criticism against applied ethics from an exactly opposite direction. They have invoked notions of morality that perceive the ethical demand as impossible to conceptualize, based on Wittgenstein's account of language and ethics. Morality, moreover, appears to be deeply subtle and personal in quandaries that applied ethics aims to study. Moreover, even if a conceptualization of ethics succeeded, the critics claim, it would be equally difficult to "apply" it in concrete situations rich with aspects. It has been, furthermore, argued that a philosopher solving moral problems that are embedded in people's personal lives is unpalatable for a serious understanding of morality, since no one can take another's place in the moral sense. Following early Wittgenstein, these critics find applied ethics too conceptual, since "the matters of the heart" or the sound of the conscience cannot be captured by conceptual thinking.[98] Critics say that a legal, technical, or scientific expert is an understandable concept, but expertise in ethics is an awkward notion. Therefore, they look with contempt upon ethical experts who are ready to solve moral problems within professions.[99] Partly because of these considerations, applied ethics has gained little respect as a philosophical subdiscipline and research area, although teaching in applied ethics is required more and more.

Continental Perspectives in Morality and Technology

In the continental, phenomenological tradition, the problem of ethics centers around Martin Heidegger's thinking, since his work opened the path for most of continental European thinking but simultaneously had a troubling effect on thinking about morality. Heidegger's thinking centers on the question of Being. After the tradition-setting phase in Greek philosophy this question was closed and forgotten, according to Heidegger, who aimed to reawaken it. The question concerns being of beings (or entities): it is crucial to understand that the Being by virtue of which beings exist is not in itself a being. Heidegger's thinking takes place as a study of this "ontological difference"—between multitude of beings and their possibility to be.

Heidegger, however, was careful to note that his philosophy had nothing to do with ethics, even when it might have outwardly appeared to. In order to account for Being, Heidegger needed to analyze carefully how a human being exists; the questioning of Being takes place in a human being as humans are the only beings in whom this question occurs. Moreover, following an ancient philosophical track, Heidegger also showed how this questioning can be appropriately carried on only by a human being who lives, and has lived, in a certain way. In *Being and Time* (1927), one central theme concerns the ways in which *Dasein* (i.e., human being analyzed phenomenologically) strives for more *eigentlich* (authentic, true, personal) existence, out of its forgetful *Verfallenheit* ("fallenness"). This could be a moral theme of personal ethics. Yet, according to Heidegger, it has nothing to with morality.[100] On the contrary, morality belongs to the *uneigentlich* mode of existence: attempting to shape one's life according to some objective or even intersubjective moral criteria is quite obviously in conflict with authenticity.[101] Sections that treat the ways in which *Dasein* may relate with others so that their authentic being is enabled contain kernels for ethical thinking that would be acceptable to Heidegger. He, however, does not develop them further.[102] Moreover, what looks like Heidegger's ethos of *Eigentlichkeit* has considerable internal problems. A central feature of *eigentlich* existence is resoluteness (*Entschlossenheit*): *Dasein*'s coming back to its ownmost possibilities of existing, rejecting *das Man* ("one," the indefinite subject of existence avoiding responsibility) and not shying away from mortality.[103] Heidegger, however, does not bring forth any criteria for the content or direction of the resolution. Thus, *Being and Time* either suppresses the moral themes or remains silent about them.[104]

Heidegger's main concern was that ethics is entangled in a metaphysics that conceals Being. Ethics as traditionally understood seems to be rules for human entities, and such thinking ignores the character of human existence and the question of being. Heidegger, however, did not attempt to describe how ethics could be thought of in an ontologically proper way.[105]

Heidegger's work, nonetheless, in an indirect way sheds light on ethics through his thinking on the problematic situation of modern culture. Heidegger examines the condition of modern culture under the rubric of nihilism.[106] For Heidegger, nihilism means a historical condition in which the highest ideals of the culture have lost their binding force on humans. The nihilistic era began dramatically, with the idea of the death of God. In brief, this idea refers to the ousting of Christian (and metaphysical) God from the culture; God ceases to occupy the role of the (nearly) unquestioned, highest good and reality. For Friedrich Nietzsche (1844–1900), who started the discussion on the death of God, this change and the nihilism that followed it were promising prospects: they provided a stage for overcoming religion and metaphysics that placed the good beyond immanent living—in the afterworld, in ideas and ideals, in submissive morals. Heidegger, however, was more ambivalent and analytic; he confined himself to reflective witnessing of the historical situation without passing judgment.[107]

In Heidegger's view, as the central notion in Nietzsche's work—he urged future philosophers to reevaluate all values—and in many strands of modern traditions, the concept of value is itself nihilistic:

> The representation of value is just as essential to the modern interpretation of that which is, as is the system. Where anything that is has become the object of representing, it first incurs in a certain manner a loss of Being. This loss is adequately perceived, if but vaguely and unclearly, and is compensated for with corresponding swiftness through the fact that we impart value to the object and to that which is interpreted as object, and that we take measure of whatever is, solely in keeping with the criterion of value, and make values themselves the goal of all activity. . . . Value is the objectification of needs as goals, wrought by a representing self-establishing within the world as picture. Value appears to be the expression of the fact that we, in our position of relationship to it, act to advance just that which is itself most valuable; and yet

> that very value is the impotent and threadbare disguise of the objectivity of whatever is, an objectivity that has become flat and devoid of background. No one dies for mere values.[108]

Simplifying somewhat, Heidegger's point is that in themselves values appear to be "objective"—as is everything that is to be counted real in a modern worldview—but for each value, its being valuable depends on the subject's choice or valuation. Consequently, values are not good in themselves; they are "good" only if they are valued by someone. In the same manner, religion, in Heidegger's reading, turns into "mere religious experience."[109] In sum, Heidegger argues that nihilism does away with the highest ideals of the Western tradition, and not only some particular ideals but the binding ideality itself.[110]

In addressing technology Heidegger was more elaborate than Wittgenstein. In *Being and Time* Heidegger drew a description of human existence in a manner that accorded an unusually significant importance to the pre-reflective, practical, productive aspects of life. In Heidegger's analysis, *Dasein* (human existence, "being there") is inescapably in the world, which means that the practical, finite, temporal viewpoint is the primary way a human being perceives reality. Theoretical abstractions, permanent truths, and the scientific perspective in general, according to Heidegger, are only secondary modifications of the primary practical and concrete way of being in the world.[111]

Many of Heidegger's later texts, after the Second World War, address more or less directly technology as an epoch-making phenomenon. In Heidegger's treatment modern technology and modern science are perceived in tight connection. The most famous and articulate text presenting Heidegger's philosophy of technology is "The Question Concerning Technology" ("*Die Frage nach Technik*"), which he delivered as a lecture in 1949 and published in an extended form in 1954. In this text Heidegger describes the rule of technology through the notion *Ge-stell* (enframing). *Ge-stell* determines the way modern people encounter reality. Technological enframing is the way of revealing truth during the technological epoch. Reality disclosed within enframing, however, becomes "a standing reserve," or resource, for technologized humanity. This development leads into a situation in which "[i]t seems as though man everywhere and always encounters only himself [his construct]." And therefore, "[t]he rule of enframing threatens man with the possibility that it could be denied to him to enter into a more original revealing and hence to experience the call of a more primary truth."[112] Simplifying somewhat, Heidegger sees it as a danger that human beings could end up perceiving the mathematical-scientific framework within which manipulative technology discloses the world as the only reality, without understanding to what extent it is a human projection.

These Heideggerian themes were developed further by a number of the 20th-century philosophers. Hannah Arendt (1906–1975), for instance, analyzed how the sphere of political activity—Aristotelian *praxis*—vanishes in modern society as technological and economic viewpoints gain hegemony. The Frankfurt School social theorists' analyses of the instrumental reason, such as Herbert Marcuse's *One-Dimensional Man*, draw on Heidegger as well. Some thinkers who understood Heidegger on his own terms, however, perceived his philosophy as the manifestation of nihilism, "historicism," and groundless relativism. These critics of Heidegger include Hans Jonas, Leo Strauss (1899–1973), and Karl Löwith (1897–1973).[113] In this study, I shall treat Jonas's ethics of technology in chapter 3 and Albert Borgmann's neo-Heideggerian analysis of everyday life in technologized societies in chapter 4. In some respects, Borgmann's analysis comes close to the themes developed by the Frankfurt School. His political thinking, however, rejects Marxism and practices serious dialogue with liberal democratic tradition, foremostly John Rawls.

After this journey through some vantage points into the groundlessness of ethics in the 20th century, the observations made can be summed up as follows. Academic philosophy has been reluctant to treat ethics seriously and has tended to regard views that present ethics as arbitrary or problematic as rational. This situation in academia reflects the wider historical situation that is characterized by scientific reason as the measure of the real; disenchantment following from this rule of science, subjectivization of valuation and meaning; and the dominance of instrumental reason over the classical understanding of politics. Without exaggeration, one could say that the contemporary era is technological, that we live in a technological form of life, or that our way of being in the world is fundamentally technological. Such overall characterizations, however, help little with the task of thinking about ethics and technology. Be the character of an era what it may, humans act and decisions are made—this is what ethics attempts to address.

HOW TO ADDRESS THE QUESTIONS

I will conclude this chapter with some reflections on the requirements for a research approach that aims at addressing the relationship between ethics and technology or at constructing an ethics of technology. As this chapter has shown, a few distinct approaches to the questions discussed have emerged during the past decades. First, we encountered the various branches of applied ethics that attempt to solve moral questions in different technologically created settings. Second, in the introduction, science, technology, and society studies were identified as programs that study technology empirically within the framework of social science. Third, throughout this study I have remarked on the ways philosophers have addressed technological advance. Thus, the philosophy of technology forms a distinct discourse relevant to ethics and technology issues.

My proposal is that a reflection on ethics and technology—and ethics of technology—needs to be informed by all three of these research programs if it aspires to be comprehensive and avoid some evident shortcomings.[114] I do think that philosophy of technology, and the philosophical tradition more generally, is the best starting point for ethics of technology, for reasons I will clarify below. Yet philosophy can benefit from contact with the other research programs.

Common to applied ethics and fields that address "the challenges" created by technological advance is that they seldom focus on those technologies from which the challenges result. Moreover, they are often somewhat regional and restricted in their approach. Although both environmental ethics and bioethics may be understood broadly as ethics of human survival, Anglo-American applied ethics most often tends to be issue centered and individualistic in character. These I perceive as shortcomings of applied ethics in general and, in particular, in the face of the ethics and technology questions. Some of this theoretical lack of depth and comprehensiveness in applied ethics is apparently a result of its history: metaethically oriented analytic ethics was unprepared for practical and normative tasks that were assigned to it after the Second World War. Applied ethics, nonetheless, possesses something that is crucial for ethics and technology discourse, as well: the perspective of moral agency.

Furthermore, philosophy of technology needs to be informed by science, technology, and society (STS) studies. STS possesses the strength of providing a broad view of technology that applied ethics lacks. Moreover, STS aims at nuanced empirical—in contrast to speculative, not really falsifiable, armchair cultural philosophy—studies on technology. This constitutes an irreplaceable contribution to ethics and technology discourse. STS, however, has little or nothing to say about ethics. In the present phase of the field's development, STS scholars are hesitant to take normative stances but lean toward the ideas of a value-free social science. This, obviously, is a deficit from the perspective of ethics. Moreover, social scientists often criticize STS for having a shallow connection to the discussion in general social theory.[115]

Philosophy of technology forms the core of the suggested interdisciplinary approach. Although philosophers have rarely addressed technology, the possible philosophical positions on technology are many and varied in starting points and styles. Carl Mitcham divides the philosophy of technology into two main traditions: engineering philosophy of technology and humanities philosophy of technology.[116] Following his division, it appears that an engineering philosophy of technology would assign the tasks of ethical reflection to applied ethics and be satisfied with that approach. Analytic philosophy of science, engineering philosophy of technology, and applied ethics appear to a large extent to share the same intellectual framework. This framework leaves the subject of ethics and science, rationality, and scientific objectification unproblematized.

A humanities philosophy of technology, which mostly exists within the continental tradition of modern philosophy, however, is in many ways superior to the approaches of applied ethics and STS.[117] While applied ethics is blind to technology as a world-making and structural phenomenon, in continental philosophy technology is most often reflected upon exactly in this comprehensive sense. Moreover, whereas applied ethics looks at human agency in a rather unproblematic manner, continental philosophers have taken it as their task to analyze the not-entirely-conscious background of conscious agency and the practices that shape the human subject. For STS the philosophical reflection on technology may provide a theoretical depth that an empirical case study–based approach often lacks. The philosophy of technology, however, has its deficits as well. In its classical forms, for instance in Heidegger, Jacques Ellul, and Lewis Mumford's thinking, the philosophy of technology is speculative and empirically unfalsifiable.[118] Furthermore, the idea of practical moral agency often tends to disappear in lofty analyses of human condition, a situation that becomes the most evident in a deterministic understanding of technological change. For the first deficit, STS is obviously the corrective supplement; for the second, a dialogue with applied ethics would be fruitful.

In brief, my remarks on the three research programs are as follows. Applied ethics has the asset of keeping moral agency open, but it is blind to technology as a force that is conditioning human agency. STS is strong in researching technology empirically as a concrete, "physical-social" phenomenon but hesitates to take normative stances. Moreover, the recent academic specialization among STS scholars has led to a narrowing of their approaches. Continental philosophy of technology can shed light on the agency that applied ethics leaves untouched. Although philosophy of technology sometimes lacks contact with the empirical realm of technology, it can best address technology in the human condition in a radical and comprehensive manner. Both applied ethics and STS, in many ways, remain within the confines of contemporary academia—narrow research approaches, political neutrality, and specificity in methodologies. Philosophical tradition, however, has had the power to assess human ways of taking up with reality, and more than that: philosophy from its earliest phases has been able to question human ways of life (Gr. *bios*) and articulate new ones.

To conclude, in this chapter I have outlined what I consider the most important questions in ethics and technology within a threefold structure that I argued to underlie those questions: (1) challenges that technology poses for ethics, (2) possible revisions needed for ethical thinking in a world altered by technology, and (3) foundational problems of ethics in modern conditions. In the end I proposed an interdisciplinary, philosophical approach as the best avenue to address the questions touched upon in this chapter. In the next chapters I will read the work of the three thinkers in light of the observations, arguments, and conclusions of this chapter.

NOTES

1. Jonathan Glover, *Humanity: A Moral History of the Twentieth Century* (2001), 64–118. Dave Grossman, *On Killing: The Psychological Cost of Learning to Kill in War and Society* (1996), 99–192.

2. Concerning ways to structure the levels of the relationship between technology and ethics, compare, for instance, Hickman, *Philosophical Tools* (2001), 3, and Jozef Keulartz, Maartje Schermer, Michiel Korthals, and Tsjalling Swierstra, "Ethics in Technological Culture: A Programmatic Proposal for a Pragmatist Approach," *Science, Technology, and Human Values* (2004).

3. Michael Davis, *Thinking Like an Engineer: Studies in the Ethics of Profession* (1998), 148. Mike W. Martin and Roland Schinzinger, *Ethics in Engineering* (2005), 91, 135–141, 245–248.

4. Rachel Carson, *Silent Spring* (1987, first published 1962).

5. Martin and Schinzinger, *Ethics in Engineering*, 230.

6. James H. Moor, "What Is Computer Ethics?" *Metaphilosophy* (1985). Deborah G. Johnson, *Computer Ethics* (2001); "Computer Ethics," *A Companion to Applied Ethics* (2003).

7. Raymond E. Spier, "Nuclear Ethics: Industrial Perspectives"; Carl Mitcham, "Nuclear Ethics: Weapons Perspectives," *Encyclopedia of Science, Technology, and Ethics* (2005).

8. Paul B. Thompson, *The Spirit of the Soil: Agriculture and Environmental Ethics* (1995).

9. Committee on Science, Engineering, and Public Policy, *On Being a Scientist: Responsible Conduct in Research* (1995). National Advisory Board on Research Ethics, *Good Scientific Practice and Procedures for Handling Misconduct and Fraud in Science* (2002). Carl Mitcham and Robert Frodeman, "New Directions on the Philosophy of Science," *Philosophy Today* (2004).

10. David W. Gill, "The Technological Blind Spot in Business Ethics," *Bulletin of Science, Technology, and Society* (1999); "Business Ethics," *Encyclopedia of Science, Technology, and Ethics* (2005). Engineering ethics, bioethics, and environmental ethics are briefly reviewed below.

11. Carl Mitcham and Helen Nissenbaum, "Technology and Ethics," *Routledge Encyclopedia of Philosophy* (1998). Daniel Callahan, "Bioethics," *Encyclopedia of Bioethics* (2003).

12. The word derives from Latin *ingenerator*, which referred to people who designed and constructed battering rams, catapults, and other devices of war. In the modern world, English John Smeaton (1724–1792) was the first engineer to call himself a "civil engineer" as his works were primarily public. Carl Mitcham, "Schools for Whistle-Blowers: Educating Ethical Engineers," *Commonweal* (1984).

13. Davis, *Thinking Like an Engineer*, 18–30. Charles E. Harris, Michael Pritchard, and Michael J. Rabins, "Engineering Ethics," *Encyclopedia of Science, Technology, and Ethics* (2005).

14. Gary Downey and Juan Lucena, "Engineering Cultures," *Science, Technology, and Society: An Encyclopedia* (2005).

15. Michael S. Pritchard, "Engineering Ethics," *A Companion to Applied Ethics* (2003). Harris et al., "Engineering Ethics." Christelle Didier and Bertrand Hériard Dubreuil, "Engineering Ethics: Europe," in *Encyclopedia of Science, Technology, and Ethics* (2005).

16. Davis, *Thinking Like an Engineer*, 16, 36, 39, 142–143. Martin and Schinzinger, *Ethics in Engineering*, 44–52, 67, 70.

17. Richard T. De George, *Business Ethics* (1999), 240–261. Cf. Mike W. Martin, "Whistle-Blowing: Professionalism, Personal Life, and Shared Responsibility for Safety in Engineering," *Business and Professional Ethics Journal* (1992). Martin and Schinzinger, *Ethics in Engineering*, 172–179. Hollywood has occasionally, for example in *Silkwood* (1983) and *Erin Brockovich* (2000), made use of the heroic potential of whistle-blowing.

18. Langdon Winner, "Citizen Virtues in a Technological Order," *Technology and the Politics of Knowledge* (1995), 74. Martin and Schinzinger, *Ethics in Engineering*, 178–179.

19. Davis, *Thinking Like an Engineer*, 85–106. Pritchard, "Engineering Ethics." Harris et al., "Engineering Ethics."

20. Radford Byerly, Jr. "Space Shuttles *Challenger* and *Columbia* Accidents," *Encyclopedia of Science, Technology, and Ethics* (2005). Davis, *Thinking Like an Engineer*, 42–44, 63–68, 73, 120–125. Harry Collins and Trevor Pinch, *The Golem at Large: What You Should Know about Technology* (1998), 30–56. Martin and Schinzinger, *Ethics in Engineering*, 106–111.

21. Harris et al., "Engineering Ethics." Donald C. Ainslie, "Principlism," *Encyclopedia of Bioethics* (2003).

22. Langdon Winner, "Engineering Ethics and Political Imagination," *Broad and Narrow Interpretations of Philosophy of Technology* (1990). Willem H. Vanderburg, "Preventive Engineering: Strategy for Dealing with Negative Social and Environmental Implications of Technology," *Journal of Professional Issues in Engineering Education and Practice* (1995).

23. Samuel C. Florman, *Blaming Technology* (1981), 162–180. Martin and Schinzinger, *Ethics in Engineering*, 67–69. Harris et al., "Engineering Ethics."

24. Roderick Frazier Nash, *The Rights of Nature: A History of Environmental Ethics* (1989), 121–160.

25. Nash, *The Rights of Nature*, 33–55. Robin Attfield, *Environmental Ethics: An Overview for the Twenty-First Century* (2003), 1–24, 37–39. Andrew Light, "Environmental Ethics," *A Companion to Applied Ethics* (2003). David Strong, "Environmental Ethics," *Encyclopedia of Science, Technology, and Ethics* (2005). Susan J. Armstrong and Richard G. Botzler (eds.), *Environmental Ethics: Divergence and Convergence* (1993) offers an ample review of environmental ethics positions.

26. Attfield, *Environmental Ethics*, 31–37. Light, "Environmental Ethics."

27. Attfield, *Environmental Ethics*, 31–58, 78–84.

28. Albert Jonsen, *The Birth of Bioethics* (1998), 3–89. Daniel Callahan, "Bioethics," *Encyclopedia of Bioethics* (2003). Margaret P. Battin, "Bioethics," *A Companion to Applied Ethics* (2003). Minttu Talosela, *Hajatelmia bioetiikasta* (2005).

29. Talosela, *Hajatelmia bioetiikasta*. Battin ("Bioethics," 297–298, 307–311) perceives a progress in bioethics from individualistic, issue-centered medical ethics toward a broader bioethics that addresses deeper theoretical questions and is capable of social criticism.

30. The discussions are vast; Bo Rothstein's *Just Institutions Matter: The Moral and Political Logic of the Universal Welfare State* (1998) and Christopher Pierson's *Beyond the Welfare State? The New Political Economy of Welfare* (2006) provide an overview.

31. Gavin Kitching, *Seeking Social Justice through Globalization: Escaping a Nationalist Perspective* (2001), 105–125. It is debatable to what extent globalization is a result of—especially information—technological development and to what extent it results from political decisions.

32. Kitching, *Seeking Social Justice through Globalization*, 299–318.

33. Carson, *Silent Spring*. Pablo A. Iannone, "Pollution," *Encyclopedia of Science, Technology, and Ethics* (2005).

34. The seriousness of a risk is a combination of quantitatively assessed risk "function" and social risk evaluation.

35. Kristin Shrader-Frechette, *Risk and Rationality: Philosophical Foundations for Public Reforms* (1991). Langdon Winner, *The Whale and the Reactor: A Search for Limits in an Age of High Technology* (1986), 138–154. S. O. Funtowicz, J. Martinez-Alier, G. Munda, and J. R. Ravetz, *Information Tools for Environmental Policy under Conditions of Complexity* (1999).

36. Brian Martin, "Technological Vulnerability," *Technology in Society* (1996). Amory B. Lovins and L. Hunter Lovins, "Brittle Times, RMI's Response," *RMI Solutions* (2001). The accident in the Fukushima nuclear power plant that followed the tsunami in Japan on March 11, 2011, is the latest and one of the most severe examples of technological vulnerability.

37. Kristin Shrader-Frechette, "Technology Assessment," *Encyclopedia of Bioethics* (1995).

38. OTA, however, was closed down by Congress in September 1995, after twenty-three years of work developing and practicing technology assessment.

39. Shrader-Frechette, "Technology Assessment."

40. Ibid.

41. Tom L. Beauchamp and Ruth R. Faden, "Informed Consent," *Encyclopedia of Bioethics* (1995). Ruth R. Faden and Tom L. Beauchamp, *A History and Theory of Informed Consent* (1986), 53–113, 235–273, 298–336.

42. Shrader-Frechette, *Risk and Rationality*. Mitcham and Nissenbaum, "Ethics and Technology." Martin and Schinzinger, *Ethics in Engineering* (88–116) develop an argument that engineering projects are social experiments. Following this line of thought, one could consistently argue that if informed consent is required for biomedical operations and experiments, then similarly it ought to be required for "social experiments."

43. In general, the public spheres for reflection on technologies and appropriate decision-making appear to be scarce. See Winner, "Citizen Virtues in a Technological Order," 73–77.

44. Johan W. Schot, "Constructive Technology Assessment," *Encyclopedia of Science, Technology, and Ethics* (2005).

45. Langdon Winner, *Autonomous Technology: Technics-out-of-Control as a Theme in Political Thought* (1977), 73–106. The ideas of technological determinism and the autonomy of technology were characteristic of "the classical philosophy of technology." The central thinkers who profess such views are Martin Heidegger, *The Question Concerning Technology and Other Essays* (1977); Jacques Ellul, *Technological Society* (1964); and Lewis Mumford, "Technics and the Nature of Man," *Philosophy and Technology: Readings in the Philosophical Problems of Technology* (1972). The basic philosophical views of technological change are instrumentalism and technological determinism. According to instrumentalism, technology is a neutral tool based on neutral scientific knowledge. Instrumentalism depicts human agents free to choose the ends that they pursue with technological means. According to technological determinism, technology is an autonomous, substantial historical force that effects social change. Technological determinism does not attribute much freedom to human agents; it rather depicts technology as setting the goals and modifying human desires. Technological determinism does not only belong to comprehensive theories. Many mainstream simplistic historical interpretations, such as "the invention of the printing press enabled the Reformation" or "the pill made sexual liberation possible," have deterministic ingredients. See Albert Borgmann, *Technology and the Character of Contemporary Life: A Philosophical Inquiry* (1984), 7–12; Larry A. Hickman, *John Dewey's Pragmatic Technology* (1990), 8–16, 140–165.

46. These discussions have revolved around Langdon Winner's (*The Whale and the Reactor*, 19–39) article "Do Artifacts Have Politics?" (first published 1980).

47. James M. Jasper's "Three Nuclear Energy Controversies" (*Controversy: Politics of Technological Decisions*, 1992) elaborates on how different political systems relate to nuclear energy. His exemplary cases are the United States, France, and Sweden. *Edge of Darkness* (1985), a miniseries written by Troy Kennedy-Martin and directed by Martin Campbell, insightfully addresses nuclear waste as a political problem.

48. Andrew Feenberg, *Questioning Technology* (1999), 90–91.

49. Winner, "Citizen Virtues in a Technological Order," 73–82. Richard E. Sclove, "Nuts and Bolts of Democracy: Democratic Theory and Technological Design," *Democracy in Technological Society: Philosophy and Technology* (1992), 139–157; "STS on Other Planets," *Visions of STS: Counterpoints in Science, Technology, and Society Studies* (2001).

50. René Descartes, *Discours de la méthode* [*Discourse on Method*] (1988), 142–143, AT VI: 62. "[P]ar laquelle, connaissant la force et actions du feu, de l'eau, de l'air, des asters, des cieux et tout les autres corps qui nous environment, aussi distinctement que nous connaissons les divers métiers de nos artisans, nous les pourrions, employer en même façon à tous les usages auxquels ils sont propres, at ainsi nous render comme maîtres et possesseurs de la

nature. Ce qui n'est pas seulement à desirer pour l'invention d'une infinite d'artifices, qui feraient qu'on jourait, sans aucune peine, des fruits de la terre et de toutes les commodités qui s'y trouvent, mais principalement aussi pour la conservation de la santé, laquelle est sans doute le premier bien et le fondement de tous les autres biens de cette vie" (Descartes, *Discours de la méthode*, 1988, 634, AT VI: 62). Earlier formulations of this idea appear in Francis Bacon's writings, for example, in the preface to *Great Instauration* (1620) and book one of the *Novum Organum*.

51. Leo Marx, "Does Improved Technology Mean Progress?" *Technology Review* (1987). Thomas Hughes, *Human-Built World: How to Think about Technology and Culture* (2004), 17–43. Carl Mitcham, *Thinking through Technology: The Path between Engineering and Philosophy* (1994), 283–289. For instance, in 1813, Thomas Jefferson wrote to John Adams on the effects of the new science and the American Revolution in Europe: "Science had liberated the ideas of those who read and reflect, and the American example had kindled feelings of right in the people. An insurrection has consequently begun, of science, talents, and courage, against rank and birth, which have fallen into contempt. . . . Science is progressive" (quoted in Marx, 1987). It is worth noting that in the late 20th and early 21st centuries the rhetoric that would emphasize technology as expanding democracy, freedom, and equality is seldom invoked. (The "inherently democratizing" character of the Internet might be the only exception.) The novelty and technical superiority of the new technology appear to be good enough arguments in favor of it.

52. Perhaps it should be noted, however, that the People's Republic of China is an example of aggressive technological development that does not fully work in tandem with better chances of political participation or self-realization in all its forms.

53. Ecological problems and the question of the good life are related in many complex ways. Two obvious connections, however, are easy to point out: first, both are creations of modern technoscientific advancements, and second, technological disengagement in personal life-world often results in a person's lack of awareness of ecological realities.

54. Critiques of mass culture, such as José Ortega y Gasset's (1883–1955) *The Revolt of the Masses* (1950) [*La rebelión de las masas*, 1929], have addressed the character of living in technologically advancing societies. Although Ortega in his book does not treat technology—as he did a few years later in *Meditacion de la technica* (1939)—he puts forth a blunt critique of "the mass-man." This human type, according to Ortega, lacks the ideals and desire to mold something excellent out of his/her life and in a self-satisfied manner scorns attempts at nobility and the service of higher causes. Herbert Marcuse (1898–1979), in *One-Dimensional Man: Studies in the Ideology of Advanced Industrial Society* (1964), analyzed the "one-dimensional rationality" that he saw as gaining hegemony in advanced nations, both communist and capitalist. He argued that through the creation of false needs, people adapt to the socially repressive industrial society. By "one-dimensional rationality" he meant the narrow conception of reason that is unable to practice critique while it is capable of instrumental adaptation. Various analyses of *inauthenticity* in the modern world come close to inquiring into the relationship between technology and the good life. Most of these analyses are indebted to Martin Heidegger's treatment of *das Man* in *Being and Time* (1962, originally published as *Sein und Zeit*, 1927), which explores the "fallen" (*verfallen*) modes of human existence. Yet Heidegger denied that his analysis was in any sense normative. The Marxist concept of *alienation* (*Entfremdung*) refers to the human condition in a technological world. Marx, however, perceived the distorted relations of production (i.e., exploitative proprietary and power relations), not technology, at the heart of the problem.

55. Borgmann, *Technology and the Character of Contemporary Life*, 124–143.

56. Carl Mitcham, Introduction, *Encyclopedia of Science, Technology, and Ethics* (2005).

57. Robert McGinn, "Technology, Demography, and the Anachronism of Traditional Rights," *Journal of Applied Philosophy* (1994). Hans Jonas, *Imperative of Responsibility: In Search of an Ethics for the Technological Age* (1984), 117–123, 136–140.

58. ". . . es also keiner Wissenschaft und Philosophie bedürfe, um zu wissen, was man zu tun habe, um ehrlich und gut, ja sogar um weise und tugendhaft zu sein" (Immanuel Kant, *The Grounding for the Metaphysics of Morals* [*Grundlegung zur Metaphysik der Sitten*], 1993, AK 404).

59. McGinn, "Technology, Demography, and the Anachronism of Traditional Rights."

60. Carl Mitcham, "Responsibility and Technology: The Expanding Relationship," *Technology and Responsibility* (1987); "Responsibility: Overview," *Encyclopedia of Science, Technology, and Ethics* (2005).

61. Mitcham, "Responsibility: Overview."

62. Mitcham, "Responsibility: Overview." Glover, *Humanity*, 89–112.

63. Martin and Schinzinger, *Ethics in Engineering*, 213–214. Jonas, *Imperative of Responsibility*, 6–8, 25–31.

64. Clark Wolf, "Intergenerational Justice," *A Companion to Applied Ethics* (2003).

65. The word "cyborg" comes from "cybernetic organism," i.e., a self-regulating organism that has both artificial and organic parts. Cyborgs have been central characters in popular science fiction films and TV series, for example in *The Six Million Dollar Man* (1973), *Star Wars* (1977), *Blade Runner* (1982), *Terminator* (1984), and *Battlestar Galactica* (2004).

66. Donna Haraway's "A Cyborg Manifesto: Science, Technology, and Socialist-Feminism in the Late Twentieth Century" (*Simians, Cyborgs, and Women: The Reinvention of Nature*, 1991, 149–182) is the central text in discussions on cyborgs.

67. Marshall McLuhan, *Understanding Media: The Extensions of Man* (1964/1994). Don Ihde, *Technology and the Lifeworld* (1990). Investigation report (A 1/2004 Y) of the accident in Konginkangas, March 19, 2004, provides a detailed account of the interpenetration of technical data, legal regulations, and attribution of responsibility. This accident, which involved a charter coach and a heavy vehicle, was the most disastrous collision on Finnish roads in history and resulted in the deaths of 23 people. The investigation board found 32 direct and indirect causal factors in the accident.

68. Larry Arnhart, "Human Nature," *Encyclopedia of Science, Technology, and Ethics* (2005).

69. Herbert Marcuse, *One-Dimensional Man* (1964), 1–83. Martin Heidegger, "Memorial Address," *Discourse on Thinking* (1966), 43–57. Ellul, *Technological Society*, 340–410.

70. Raymond Kurzweil, *The Age of Spiritual Machines: When Computers Exceed Human Intelligence* (1999). See Julian Savulescu and Nick Bostrom (eds.), *Human Enhancement* (2011).

71. Francis Fukuyama, *Our Post-Human Future* (2002), 3–104. Jürgen Habermas, *The Future of Human Nature* (2003), 16–100.

72. Carl Mitcham and Adam Briggle, "Humanization and Dehumanization," *Encyclopedia of Science, Technology, and Ethics* (2005). Carl Mitcham, "Philosophy of Technology," *Guide to Culture of Science, Technology, and Medicine* (1984), 339–344.

73. Aristotle, *Nicomachean Ethics*, 1139a26–b9; *Metaphysics*, 1025b18–26, 1064a10–18.

74. Borgmann, *Technology and the Character of Contemporary Life*, 40–48, 78–85, 114–143; *Crossing the Postmodern Divide* (1992), 110; "The Moral Significance of the Material Culture," *Technology and the Politics of Knowledge* (1995); *Real American Ethics: Taking Responsibility for Our Country* (2006), 175–188.

75. Arguments for the political legitimation of technology on the basis of scientific knowledge are becoming rarer. Often, the promise of economic gain is the primary and sufficient premise in arguments legitimating investments in science and technology. The Science and Technology Policy Council of Finland (2006) summarizes goals for Finnish policy as follows: "The aim of development measures is to: (1) promote the overall functionality of the innovation system and the system's ability to renew itself, (2) enhance the knowledge base, (3) improve the quality and targeting of research, (4) promote the adaptation and commercialisation of research results, and (5) secure adequate economic prerequisites for the activities. The continuous development of human resources ensures top-quality competence for the future as well."

76. Harry Collins and Trevor Pinch's *The Golem: What You Should Know about Science* (1998, esp. 1–5, 27–56, 139–150) presents in an accessible form the viewpoint and some results of the recent sociology of science. This social-constructivist approach takes seriously the character of science as human action, in the same way that art, religion, politics, and education are human action and have been the object for sociological studies for a long time. Bruno Latour's *We Have Never Been Modern* (1993, esp. 13–48, 130–145) treats the same topic in a more philosophical style. Science, technology, and society (STS) studies have, after the activist

phase in the 1960s and '70s, to a large extent succumbed to the value-neutral scientific ethos. This has happened in parallel with the professionalization of the field. See Langdon Winner's "Upon Opening the Black Box and Finding It Empty: Social Constructivism and the Philosophy of Technology" and Wiebe Bijker's "Do Not Despair: There Is Life after Constructivism" (*Science, Technology, and Human Values*, 1993) for the discussion on normativity and STS.

77. This allegation obviously depends on the definitions. Some might find a number of candidates: utilitarianism, deep ecology, and other forms of environmental thinking, for example.

78. Karl-Otto Apel, "The Problem of a Macroethic of Responsibility to the Future in the Crisis of Technological Civilization: An Attempt to Come to Terms with Hans Jonas's 'Principle of Responsibility,'" *Man and World* (1987), 22. Elsewhere, Apel points out that the rational (and in an extreme case any) foundation of ethics is also threatened from an opposite direction: he refers to "a new wave of historicism and relativism, coupled with a positivism of power and an irrationalism that rely upon Nietzsche as their authority" (13).

79. Max Weber, "Science as a Vocation," *From Max Weber: Essays in Sociology* (1948), 139ff.

80. John Passmore, "Logical Positivism," *The Encyclopedia of Philosophy* (1967). Alfred J. Ayer (*Language, Truth and Logic*, 1936, 15–39, 149–152) and Rudolf Carnap especially professed this view.

81. Ayer, *Language, Truth and Logic*, 149–150.

82. Richard B. Brandt, "Emotive Theory of Ethics," *The Encyclopedia of Philosophy* (1967). Ayer, *Language, Truth and Logic*, 149–170.

83. Ayer, *Language, Truth and Logic*, 158–160.

84. Michael L. Corrado, *The Analytic Tradition in Philosophy* (1975), 90–110. G. E. M. Anscombe, "Modern Moral Philosophy," *Ethics, Religion and Politics: The Collected Philosophical Papers*, Vol. III (1981), 26–42.

85. Ludwig Wittgenstein, *Tractatus Logico-Philosophicus* (1922), 183. "6.4. Alle Sätze sind gleichwertig. 6.4.1. Der Sinn der Welt muss ausserhalb ihrer liegen. In der Welt ist alles wie es ist und geschieht alles wie es geschieht; es gibt *in* ihr keinen Wert—und wenn es ihn gäbe, so hätte er keinen Wert. Wenn es einen Wert gibt, der Wert hat, so muss er ausserhalb alles Geschehens und So-Seins liegen. Denn alles Geschehen und So-Sein ist zufällig. Was es nicht-zufällig macht, kann nicht in der Welt liegen, denn sonst wäre dies wieder zufällig. Es muss ausserhalb der Welt liegen. 6.4.2. Darum kann es auch keine Sätze der Ethik geben. Sätze können nichts Höheres ausdrücken. 6.4.2.1. Es ist klar, dass sich die Ethik nicht aussprechen lässt. Ethik ist transscendental. (Ethik und Aesthetik sind Eins.)" (*Tractatus Logico-Philosophicus*, 182). Alfred Nordmann, *Wittgenstein's Tractatus: An Introduction* (2005), 46–48, 55–57, 60.

86. Ludwig Wittgenstein, "A Lecture on Ethics," *Philosophical Review* (1965), 11–12.

87. Nordmann, *Wittgenstein's Tractatus*, 29–20, 128–131, 178–184. David G. Stern, *Wittgenstein's Philosophical Investigations: An Introduction* (2002), 19, 75–78.

88. Ludwig Wittgenstein, *Philosophical Investigations*, §§ 1–25, 68ff, 83ff. Stern, *Wittgenstein's Philosophical Investigations*, 87–99.

89. Wittgenstein, *Philosophical Investigations*, §§ 19ff, 34. Stern, *Wittgenstein's Philosophical Investigations*, 160–169.

90. For later developments among Wittgensteinians see, e.g., Stephen Mulhall, "Ethics in the Light of Wittgenstein," *Philosophical Papers* (2002).

91. Ludwig Wittgenstein, *Culture and Value* (1980). "Die apokalyptische Ansicht der Welt ist eigentlich die, daß sich die Dinge *nicht* wiederholen. Es ist z. B. nicht unsinnig, zu glauben, daß das wissenschaftliche und technische Zeitalter der Anfang vom Ende der Menschheit ist; daß die Idee vom großen Fortschrift eine Verblendung ist, wie auch von der endlichen Erkenntnis der Wahrheit; daß an der wissenschaftlichen Erkenntnis nichts gutes oder Wünschenswertes ist und daß die Menschheit, die nach ihr strebt, in eine Falle läuft. Es ist durchaus nicht klar, daß dies nicht so ist" (56; see also, e.g., 5–7, 63). In "Wittgenstein in Relation to His Times" (1982), G. H. von Wright reflects on Wittgenstein's "conservatism" and its place in history of

thought. For common themes in Wittgenstein's and von Wright's reflections on technology, see Topi Heikkerö, "The Fate of Western Civilization: G. H. von Wright's Reflections on Science, Technology and Global Society," *Bulletin of Science, Technology, and Society* (2004).

92. Another possible, narrower way to apply the notion would be to capture technologies as forms of life. Langdon Winner, in "Technologies as Forms of Life" (*The Whale and the Reactor*, 3–18), makes use of Wittgenstein's concept in this vein. The recent social impact of information technologies could be seen as an instance of technologies as forms of life.

93. Cf. Henry Adams, *The Education of Henry Adams* (1946).

94. Stephen Darwall, "Ethical Theory," *Encyclopedia of Philosophy: Supplement* (1996). Lars Hertzberg, "Was Wittgenstein a Moral Philosopher?" *Studia Theologica* (1997), 150–153. Corrado, *The Analytic Tradition in Philosophy* (1975), 90–110. G. E. M. Anscombe's and Philippa Foot's work on virtue ethics from the late 1950s onward is a minor exception.

95. Brenda Almond, "Introduction: Ethical Theory and Ethical Practice," *Introducing Applied Ethics* (1995). Tom L. Beauchamp, "The Nature of Applied Ethics," *A Companion to Applied Ethics* (2003), 1–3. Ruth Chadwick, "Applied Ethics," *Encyclopedia of Science, Technology, and Ethics* (2005).

96. Alasdair MacIntyre, "Does Applied Ethics Rest on a Mistake?" *After Virtue: A Study in Moral Theory* (1984). Lars Hertzberg, "Moral Escapism and Applied Ethics," *Philosophical Papers* (2002); "Was Wittgenstein a Moral Philosopher?" 153–154. Beauchamp, "The Nature of Applied Ethics," 2–3, 12–14.

97. Donald C. Ainslie, "Principlism." Tom L. Beauchamp and James F. Childress, *Principles of Biomedical Ethics*, 5th Edition (2001). Darwall, "Ethical Theory," 17–18.

98. Of this Heidegger provides an articulate elaboration in his treatment of the call of conscience (*Ruf des Gewissens*). The call of conscience says nothing about the affairs of the world. In it, however, *Dasein* calls itself. The silent, empty call draws *Dasein* toward authentic (*eigentlich*) being, to keep distance from the ways of *das Man* (*Being and Time*, §§54–60). At least roughly, Heidegger's intention is the same as that of Wittgenstein and his followers: matters of conscience are quiet, not publicly shared, and without detailed instructions for action. Hence an ethical expert who publicly articulates moral guidelines is in many respects problematic for this way of understanding conscience.

99. Hertzberg, "Was Wittgenstein a Moral Philosopher?" 153–154; "Moral Escapism and Applied Ethics." This theme, in fact, is as old as philosophy itself. The analysis and critique of the sophists—"wise men," teachers of virtue and oratory—is a central theme in Plato's dialogues. See, e.g., *Protagoras*, *Meno*, *Gorgias*, and *The Sophist*. Plato's points include the following: first, it is dubious that the sophists take money for their teachings; second, most probably they do not have true knowledge about the things they teach; third, sophists appear to corrupt innocent citizens with their teachings; fourth, it is awkward that sophists aim to teach people things that should be part of everybody's basic education; and fifth, sophistry is counterfeit currency pretending to be real philosophy. Philosophers by contrast do not ask for money, take the pursuit of real knowledge seriously, acknowledge the limitations of their discourses, and do not seek publicity or fame but practice dialectics in small circles of friends.

100. Heidegger, *Being and Time*, §§ 34 B, 58, 59.

101. Ibid., §§ 5, 59, 63, 77.

102. Heidegger, *Being and Time*, §§ 26, 29, 36, 40, 41. Lawrence Vogel, "Hans Jonas's Diagnosis of Nihilism: The Case of Heidegger," *International Journal of Philosophical Studies* (1995).

103. Heidegger, *Being and Time*, §§ 62. Graham Harman, *Heidegger Explained: From Phenomenon to Thing* (2007), 73–74, 93.

104. Heidegger's view of the possibility and the need for thinking about ethics is debatable. In my view, he is evasive when it comes to ethics—especially in the context of his own practical understanding of existence. Albert Borgmann, in "Heidegger and Ethics beyond the Call of Duty" (*Appropriating Heidegger*, 2000), however, gives a more charitable reading of Heidegger's stance on ethics. According to Borgmann, Heidegger's ontological project is indirectly ethical.

105. "If the name 'ethics,' in keeping with the basic meaning of the word *εθος*, should now say that ethics ponders on the abode of the human being, then that thinking which thinks the truth of being as the primordial element of the human being, then the human being, as one who eksists, is in itself originary ethics. However, this thinking is not ethics in the first instance because it is ontology. For ontology always thinks solely the being (*ὄν*) in its being." "Soll nun gemäß der Grundbedeutung des Wortes *εθος* der Name Ethik dies sagen, daß sie den Aufenthalt des Menschen bedenkt, dann ist dasjenige Denken, das die Wahrheit des Seins als das anfängliche Element des Menschen als eines eksistierenden denkt, in sich schon die ursprüngliche Ethik. Dieses Denken ist aber dann auch nicht erst Ethik, weil es Ontologie ist. Denn die Ontologie denkt immer nur das Seiende (*ὄν*) in seinem Sein" (Heidegger, "Letter on 'Humanism'" [*Brief über den "Humanismus"*], *Pathmarks*, 1998, 271; *Wegmarken*, 1976, 356–357).

106. "Nihilism" comes from Latin *nihil*, nothing. Nihilism refers to doctrines that profess views of moral skepticism, atheism, meaninglessness of life, baselessness of morals, and absence of value in everything. In this section, only Heidegger's view of nihilism as the condition of the modern world is discussed. The term "nihilism" first appeared in Ivan Turgenev's novel *Fathers and Sons* (1862). Bazarov, one of the characters, is a nihilist, which means that he does not believe anything that is not proven true. This position obviously draws on science against more traditional religion and morality. Bazarov, in fact, is a student of medicine and practices science (collects samples, does dissections) most of the time. Nihilism acquired a special meaning and role for the Russian intelligentsia. Fyodor Dostoevsky's *Demons* (1872) and *Notes from Underground* (1864) present deep treatments of nihilists and nihilism. In *Crime and Punishment* (1866) Raskolnikov is not exactly a nihilist, but the nihilistic stance in Western culture is *sine qua non* for Raskolnikov's position that ordinary morality does not apply to the strong and gifted.

107. Friedrich Nietzsche, *The Gay Science: With Prelude in Rhymes and an Appendix of Songs* (1974), §§ 108, 125, 343. Heidegger, "Word of Nietzsche 'God is Dead,'" *The Question Concerning Technology*, 53–112. Hubert Dreyfus, "Heidegger on the Connection between Nihilism, Art, Technology, and Politics," *The Cambridge Companion to Heidegger* (1993), 290–297.

108. Martin Heidegger, "The Age of the World Picture" (a lecture given in 1938, published in *The Question Concerning Technology and Other Essays*, 142). "Gleich wesentlich wie das System ist für die neuzeitliche Auslegung des Seienden die Vorstellung des Wertes. Erst wo das Seiende zum Gegenstand des Vor-stellens geworden ist, geht das Seiende in gewisser Weise des Seins verlustig. Dieser Verlust wird unklar und sicher genug gespürt und entsprechend schnell dadurch ersetzt, daß man dem Gegenstand und dem so ausgelegten Seienden einen Wert zuspricht und überhaupt das Seiende nach Werten bemißt und die Werte selbst zum Ziel alles Tuns und Treibens macht. . . . Der Wert ist die Vergegenständlichung der Bedürfnisziele des vorstellenden Sicheinrichtens in der Welt als dem Bild. Der Wert scheint auszudrücken, daß man in der Bezugstellung zu ihm eben das Wertvollste selbst betreibt, und dennoch ist gerade der Wert kraftlose und fadenscheinige Verhüllung der platt und hintergrundlos gewordenen Gegenständlichkeit des Seienden. Keiner stirbt für bloße Werte" (Heidegger, *Gesamtausgabe, Band 5: Holzwege*, 2003, 101–102).

109. Heidegger, "The Age of the World Picture," 117.

110. Heidegger, "Word of Nietzsche 'God is Dead,'" 66ff, 105. Dreyfus, "Heidegger on the Connection between Nihilism, Art, Technology, and Politics," 290–297.

111. *Being and Time*, e.g., chapters 9–10, 12, 15–16. Harman, *Heidegger Explained*, 60–70.

112. Heidegger, "Question Concerning Technology," *The Question Concerning Technology and Other Essays*, 19ff. Mitcham, *Thinking through Technology*, 49–55. Harman, *Heidegger Explained*, 127–141.

113. Hannah Arendt, *The Human Condition* (1958). Leo Strauss, *The Rebirth of Classical Political Rationalism: An Introduction to the Thought of Leo Strauss* (1989), 27–48. Marcuse, *One-Dimensional Man*. Richard Wolin, *Heidegger's Children: Hannah Arendt, Karl Löwith, Hans Jonas, and Herbert Marcuse* (2001). Jonas's criticism on Heidegger will be introduced in some detail in the next chapter.

114. Jozef Keulartz et al., in "Ethics in Technological Culture," 3–14, frame the academic scene quite similarly as I do here. Their argument is that a pragmatist approach would be the best way to address the lacunae in other research programs.

115. See, e.g., Steve Fuller, "Is STS Truly Revolutionary or Merely Revolting?" *Science Studies* (2005).

116. Mitcham, *Thinking through Technology*, 19–61.

117. Here, reference is broad and rough. I refer to such thinkers as Edmund Husserl in his *Die Krisis*, Max Scheler, José Ortega y Gasset, Lewis Mumford, Hannah Arendt, Herbert Marcuse, Jacques Ellul, and Michel Foucault.

118. The development of establishing distance from "the classical philosophy of technology" has been under way for quite some time: philosophy of technology has become more empirically informed. This is the main theme of the collection of essays by Dutch philosophers of technology, *American Philosophy of Technology: The Empirical Turn* (2001). The authors argue that there has been "an empirical turn" in philosophy of technology.

Chapter Three

Hans Jonas

Technology and Responsibility

In at least seemingly rational terms one could well imagine the following point of view:

> Talk about our duties for future generations is currently common. My choices in life can be made to look suspicious in light of this ecological-moral perspective. The common, politically correct knowledge says that I am obliged to promote the possibility of existence for future generations. Despite the niceness of such an idea, I find in this thinking at least two curious features. First of all, very few people actually appear to live by it, although in their speech they most often subscribe to it. Second, I find no foundation for such an obligation. I can understand that people who have descendants can reasonably be thought to have made some implicit contract with future generations. But what about those of us who have refrained from procreation? Is there some treaty between us and the coming generations? I see none.[1]

The above imaginary skeptic refers to crucial problems that were recognized in the preceding chapter. The chapter revealed the need to address both the adequacy of ethical theories as tools in the present technological world and their grounding in intellectual and cultural sense: it is not clear how to articulate our duties to future generations and in what sense these duties might be binding for us. I argued that although conventional moral thinking, applied ethics in particular, provides tools for thinking about ethics in technology, mainstream thinking remains somewhat naïve in taking many of its preconditions for granted. Moreover, it is ordinarily assumed that technological transformations in society have led to few or no alterations in the conditions of a moral subject. Hans Jonas's work that is under scrutiny in this chapter brings some correctives to the lacunae faced in the earlier chapter. The core themes of this chapter are the technologically altered human action and ethic requisite for this new kind of action, the grounding of this ethic, and inclusion of future generations in ethical considerations.

The chapter opens with a brief biographical note on Jonas. Then I move on to describe Jonas's analysis of humankind in modernity, which is a diagnosis of what Jonas perceived as a dire crisis. Then follows the description of Jonas's work in search of an ontological grounding for ethics, after which I introduce Jonas's ethics of responsibility, his attempt to put forth a solution for the crisis of humanity in modern technoscientific conditions. The chapter closes with a critical assessment of Jonas's project.

HANS JONAS AND THE PREDICAMENT OF MODERNITY

Hans Jonas was one of Martin Heidegger's influential Jewish students, alongside Hannah Arendt and Emmanuel Levinas (1906–1995). Jonas finished his dissertation on the concept of Gnosticism under Heidegger at Marburg in 1928 and published his first book on Augustine and the Pauline problem of freedom in 1930. Heidegger's character as a teacher and his philosophical breakthroughs that resulted in *Being and Time* (*Sein und Zeit*, 1927) had an impact on Jonas.[2] After his teacher's infamous rectoral address in 1933, in which Heidegger expressed his support for the National Socialist Party, Jonas, however, took a critical view of Heidegger's thinking.[3] Jonas perceived an inherent connection between the thinker and the politically lapsing university professor.[4] During his lifetime Jonas articulated this critique well enough to become known as "the philosopher who took on Heidegger."[5] Jonas had tragic personal reasons for taking the Nazi connections seriously: his mother was killed in Auschwitz. Jonas himself emigrated to Britain in 1933 and to Palestine in 1935, where he worked in publishing and as a part-time teacher in the Hebrew University. Jonas had been a Zionist since his youth, and in Palestine he actively recruited Jewish men to join the war effort against Nazi Germany. In the Second World War, he fought against the Germans as an anti-aircraft artillery soldier in the Jewish Brigade of the British 8th Army. As a German citizen, Jonas volunteered for the service and refused to accept office work but wanted to do battle. He was stationed at the Italian front. After the Second World War he participated in the Israeli War of Independence.[6] Finally, in 1949, he started his postwar career at McGill University, in Montreal. In 1955 he found his academic home at the New School for Social Research, in New York City. He spent the rest of his academic career at the New School.[7]

Jonas's philosophical career appears to have had three distinct phases. In the 1920s, under Heidegger and Rudolf Bultmann (1884–1976), he started the formidable project of phenomenologically analyzing the Gnostic religion. This work resulted in *The Gnostic Religion* (1958), from the larger German original work, *Gnosis und die Spätantiker Geist* I (1934) and II (1954). Another period in Jonas's work, starting in the 1950s, revolved around the concept of life and rethinking biology. He collected the essays related to this theme in *The Phenomenon of Life: Toward a Philosophical Biology* (1966). During the third phase, which most strongly started in the 1970s, he was pressed by the need for a new kind of ethics appropriate for the technological era. The central book from this period is *The Imperative of Responsibility: In Search of an Ethics for the Technological Age* (1984), a translation of *Das Prinzip Verantwortung. Versuch einer Ethik für die technologische Zivilization* (1979). Besides its intellectual originality, *Das Prinzip Verantwortung*, which was the first book in decades that Jonas wrote first in German, and essays that followed it (collected in *Technik, Medizin und Ethik*, 1985) had a strong impact on German public life. Jonas influenced especially the German Green movement. In 1987 he received the *Friedenspreis*, the peace prize awarded by the German Book Trade.[8]

In addition to these three facets of his philosopher's career, Jonas continually wrote essays in Jewish theology. In them the postholocaust questions were central. Jonas had little patience with any attempts at theodicy. Jonas's notion of God, which he admitted to be "heretical," was of a good and caring deity—but not all-powerful. Jonas's God is dependent on humans: humans should not wait for help from God but help Him stop evil in the world. In the context of his philosophical endeavors, Jonas emphasized that in modern cultural conditions ethics cannot be grounded in religion but, rather, the basis for ethics should be sought in nature.[9]

Beneath the diverse philosophical pursuits in Jonas's work lies a common theme: reflections on the predicament of humanity in modernity. On the concrete and practical level this predicament consists of the disproportion between the technological powers of humans and the ability to live with these powers. Jonas especially emphasizes two aspects of technological powers: ways in which humans have subdued the nature in their service and the perspectives opening in technological modification of humanity itself—modern biotechnology. The first has resulted in the threat of ecological disaster and nuclear holocaust; the second may radically redefine what it means to be a human being. The lack of ability to live with increased human powers, in Jonas's work, refers to the absence of a proper morality within which to think about and judge the choices, especially those in public policy, related to these enhanced powers. The absence of morality is manifested in two major ways: first, ethics as our traditions know it does not acknowledge the new issues created by technology, and second, the intellectual conditions of Western culture have become such that no proper foundation or grounding for ethical thinking exists.[10]

Concerning the lack of applicability of traditional ethics in a technological world, Jonas's arguments reflect in part the same viewpoints that were elaborated on in the previous chapter. Jonas's central claim is that the human condition has so been altered in the technological modernization process that moral thinking, which concentrates on causally, spatially, and temporally close connections—as traditional ethics does—is inadequate to the task of addressing human action enhanced by technology. The most concrete deficiency in conventional ethical considerations is the omission of the living conditions for posterity. This was no pressing problem prior the second half of the 20th century, as humankind had no real means to alter the living conditions for posterity but was more or less at the mercy of nature. Therefore, the issue was not thematized in ethics. Another dated presupposition is the traditional idea of fixed human nature. In the modern world it is at least thinkable that in the future, humans will not be always and everywhere the same. Gene technology, medicines for the psyche, and the possibilities for behavioral control, according to Jonas, exhibit possible avenues for altering human nature. Based on these observations, Jonas puts forth his demand for revisions in ethics.[11]

Revising ethics assumes that moral thinking is meaningful and somehow binding for those who share it. Jonas, however, finds this meaningfulness of moral thinking to a large extent missing in the modern world:

> And here is where I come to a standstill, where we all come to a standstill. For the very same movement which put us in possession of the powers that have now to be regulated by norms—the movement of modern knowledge called science—has by a necessary complementarity eroded the foundations from which the norms could be derived; it has destroyed the very idea of norm as such. Not, fortunately, the feeling for norm and even for particular norms. But this feeling becomes uncertain of itself when contradicted by alleged knowledge or at least denied all support by it.[12]

Here Jonas is referring to scientific neutrality, or objectivity—the value-free (*wertfrei*) science. Jonas perceives the development of modern natural science, from Nicolaus Copernicus (1473–1543) and Galileo Galilei (1562–1642) onward, as a history of human estrangement from nature. This is obviously a complex history. The main points here, however, are the following. First, the modern mathematized notion of nature does away with all value in nature. In contrast, premodern natural philosophical systems that made use of teleological explanation understood nature as goal oriented and thus involving values in itself. Second, the mathematical picture of nature necessarily coexists with a radical chasm between human mind and nature outside of it. If nature is understood purely through mathematical functions, it becomes a mere quantitative phenomenon. In contrast, human existence, in its inwardness and conscious faculties, is essentially something other than quantified reality outside it.[13]

Both of the above points were for the first time cogently articulated in René Descartes's philosophy.[14] It established the firm distinction between thinking substance, *res cogitans*, and extensional substance, *res extensa*. For the background of the modern situation, Jonas provides a broad historical sketch of human understanding of reality. He perceives early human cultures as essentially animistic, or panpsychic: early humans saw life and spirit everywhere in nature. In Western culture, Greek philosophy meant the first emergence of dualism. Dualism achieved its ripest and most extreme formulation in Descartes's philosophy. The trajectory from Descartes's modern dualism unfolded as a competition between idealist and materialistic monism, both of which positions have seen consistent formulations in modern philosophy. Mainly because of the triumphs of science and technology, post-dualistic, materialistic monism has the upper hand against its various idealist rivals.[15]

Materialistic monism obviously claims not to be a dualistic ontology. In concrete everyday life, materialism, nonetheless, leads to a dualistic situation in which the true reality consists of matter but has human life as a bizarre, epiphenomenal outgrowth. Meaning and intentionality at work in everyday life and human praxis in general still exist in an unreduced form, although materialism does not grant them the status of being real.

An analysis of dualism, of which the Cartesian type is just one instance, forms the basis of Jonas's argumentation concerning the predicament of modernity—and, in fact, concerning the whole of Western culture. In his essay "Gnosticism, Existentialism, and Nihilism,"[16] Jonas points out parallels between ancient Gnosticism and modern existential philosophy and depicts them as extreme manifestations of the Western tendency toward dualistic nihilism. In his comparisons, Gnosticism works as a heuristic interpretative tool for understanding contemporary phenomena.

Drawing on his extensive knowledge of Gnosticism in late antiquity, Jonas highlights the features of Gnostic dualism.[17] In the Gnostic doctrines the world separates the human being from God. Thus the world is conceived as hostile, bad, not as orderly cosmos in the manner of the mainstream ancient thought. The world lacks good order, since its creation was flawed; the world was made by no real God but an ignorant, incompetent demiurge. Human body and soul (*psyche*) belong to the bad world, but the human being's true identity is spirit (*pneuma*) that is trapped in the body. This entails antinomianism: since the world is bad, its laws have command only over the body and soul, whereas *pneuma* is free from them. This position is in stark contrast with Stoicism, in which the world (*cosmos*) and its laws are considered good, since they represent order (*eunomia*) according to reason (*logos*). Therefore, the way to freedom for humans, in Stoicism, is "playing one's part well in the play of life." The Stoics both followed concrete rules and believed in a metaphysical moral law in nature. Moreover, the Stoics read the worldly happenings in light of benevolent providence (*pronoia*). The Gnostics, to the contrary, perceived the mundane goings-on as works of the brute fate (*heimarmene*). The way out of the hellish world, for Gnostics, opened out of the feeling of dread at being-in-the-world. Dread leads to deepening alienation from the world, which should culminate in the reception of gnosis (knowledge about the reason of the bondage, the human's true identity, and the way out of the world) from the savior that comes from outside of the world.[18] The divine opposite of the world Jonas describes as follows:

> As the totally other, alien, and unknown, the Gnostic God has more the *nihil* than of the *ens* in his concept. For all the purposes of man's relation to the reality that surrounds him this hidden God is a negative term; no law emanates from him—none for nature, and thus none for human action as part of natural order. His only relation to the world is the negative one of saving from the world. Antinomianism follows naturally, even if not inevitably, from these premises.[19]

Jonas argues that modern existentialism is in many essential respects similar to Gnosticism. The crux of the argument is that existentialism, especially Heidegger's philosophy, does not describe human condition as such but is a historically conditioned intellectual phenomenon, in which ingredients surface that are common to Gnosticism. In its details, the argument consists of the following. Existentialism preserves the Cartesian dualism and in many respects even radicalizes it. If the world in Gnosticism is hostile, in existentialism it is just indifferent. The distinction between authentic and inauthentic[20] existence is analogical to the *pneuma-psyche* distinction. Antinomianism manifests an evident common feature in the two movements.[21]

Accusing Heidegger of dualism[22] may sound surprising. In Jonas's sense, however, Heidegger's philosophy is clearly dualistic. Although Heidegger analyzes the temporality and finitude of human existence, he neglects corporeality, as well as ethics and other people in a positive sense. Moreover, *Dasein*'s *Geworfenheit*, "having been thrown" into the world,[23] characterizes the broken relationship between existence and an indifferent exterior reality. Furthermore, *Dasein*'s fundamental attunement is anxiety (*Angst*),[24] one feature of which is *Unheimlichkeit*, literally "unhomeliness," not being at home in the world, uncanniness. In fact, facing *Angst* opens the way toward more true, or "ownmost" (*eigentlich*), existence, with which comparisons with the Gnostic dread are easily made.[25] Moreover, the Heideggerian theme of embracing one's destiny (*Schicksal*) resembles Gnostic ideas of *heimarmene*.[26] Finally, the role of death—*Dasein*'s *nihil*—as *Dasein*'s most own possibility is open for comparisons with the Gnostic notion of the deity.[27]

The antinomianism common to both Gnosticism and existentialism is one aspect of the nihilism that arises from dualism. In modern conditions the good-in-itself makes room for value that is a function of the will:

> Meaning is no longer found but is 'conferred.' Values are no longer beheld in the vision of objective reality, but are posited as feats of valuation. As functions of the will, ends are solely my own creation. Will replaces vision; temporality of the act ousts the eternity of the 'good in itself.' This is the Nietzschean phase of the situation in which the European nihilism breaks the surface. Now man is alone with himself.[28]

Jonas argues that under modern cultural conditions, ethics becomes purely a matter of human subjective consciousness that projects value judgments on a morally neutral external reality. Modern scientific materialism and existential subjectivism are, as it were, counterparts in the dialectics of the modern worldview. According to Jonas, the subjective existential basis is too weak for ethics, since it leads either to arbitrary value content or to a contentless, formal quest for authenticity and commitment. Of the formal imperative of commitment, Heidegger again serves as the most philosophically refined example. According to Jonas, Heidegger's "resoluteness" (*Entschlossenheit*) includes no standard for evaluating the content or goal of the resolution.[29]

What is common to Gnosticism—an exception among the ancient worldviews—and the modern intellectual condition in general is the absence of cosmos with *logos*.[30] The nature revealed by modern science has order, a mathematical structure of lawful connections, but this order includes no meaning and excludes good. Thus, good turns into "value" that is posited by fiat—human or divine. In the post-Christian era, values depend merely on the human will.[31]

If natural science and existential philosophy fail to give a meaningful account of human good, one might expect that the social sciences that study human social phenomena would provide answers for "value questions." This, however, is a vain hope. As sciences, social sciences aim at following the methods of triumphant natural sciences and necessarily reify their research objects and try to explain their behavior in light of causal laws. Social science may thus provide explanations of patterns in valuations, why in certain communities or groups certain things are valued. It does not, however, address questions about the adequacy of valuations, to what extent things valued are truly worth valuing. In short, social science aims at explaining human society as part of nature. Therefore, humans as objects of social science are clearly different from the humans that practice science. In the lecture "Practical Uses of Theory," Jonas summarizes:

> Thus man-the-knower apprehends man-*qua*-lower-than-himself and in doing so achieves knowledge of man-*qua*-lower-than-man, since all scientific theory is of things lower than man the knower.[32]

In sum, Jonas perceives an unfruitful dichotomy in the intellectual life of the 20th century. On the one hand, the reigning natural scientific way of describing reality attempts to reduce everything, including organic life and conscious human mind, to quantifiable matter. This "program" produces exact and valid information on its objects but makes reality unintelligible for living, conscious, and moral beings. On the other hand, a rival position, which currently is an underdog, exists: the subjective or the idealist way to understand human life and morality. Existential philosophies exhibited its deepest and most sophisticated variations in the 20th century. Existential thinking takes human consciousness and subjective inner life seriously and in its strongest forms reads natural science as human pursuit, only a secondary derivative to the practical being-in-the-world of human existence. Moreover, science is understood as conditioned by historical transformations in the human world. Existential thinking, however, leads to another extreme concerning organic life and morality as it becomes trapped in human inwardness. It still reproduces the Cartesian dualism by overemphasizing human consciousness.

ETHICS AND ONTOLOGY

As we noticed above, in dualistic ontology an insurmountable gap exists between human consciousness, whether Platonistic soul, Gnostic spirit, Cartesian *res cogitans*, or Heideggerian *Dasein*, and external reality, whether the visible world, *res extensa*, or *die Welt*. In addition to nihilistic value subjectivism,[33] dualism is problematic in another respect: within the dualistic metaphysical framework organic life becomes nearly incomprehensible. Dualistic philosophies fail to explain why human beings have to eat. As *res cogitans* is purely immaterial and *res extensa* resembles a mechanical machine, which one of them needs food? Organic processes such as metabolism remain mysteries for a dualistic philosopher. Modern science, in practice, ends up sharing the presuppositions of dualistic philosophies. Admittedly, Darwinism has shaped modern anthropology more than philosophical arguments, but Darwinism explains no better than does a dualistic philosophy the essence of organic life.[34] In short, Jonas perceives that modern science attempts to explain life through reduction to dead matter. Although this may seem an exaggeration, it is not an unreasonable analysis when one considers the wide acceptance of overall naturalism and ideas of unified science and reductionism among scientists as well as certain analytic philosophers.[35]

> Accordingly, it is the existence of life within a mechanical universe which now calls for explanation, and explanation has to be in terms of the lifeless. Left over as a borderline case in the homogeneous physical world-view, life has to be accounted for by the terms of that view. Quantitatively infinitesimal in the immensity of cosmic matter, qualitatively an exception from the rule of its properties, cognitively the unexplained in the general plainness of physical things, it has become the stumbling block of theory. That there is life at all, and how such a thing is possible in a world of mere matter, is now a problem posed to thought. The very fact that we have nowadays to deal with the theoretical problem of life, instead of the problem of death, testifies to the status of death as the natural and intelligible condition.[36]

Ontology of Life

Against this background, Jonas in *The Phenomenon of Life* embarks upon his bold metaphysical attempt to describe and explain life in a manner that would bridge the gap between arbitrary subjectivity and neutral nature. His methodology in this pursuit already implies this bridging.[37] How can we know about life in (other) living beings? Through ourselves, based on our own experience, since we are living beings ourselves, replies Jonas. He remarks that reductionist explanations of life are absurd, as they try to understand higher forms of life through the lower ones. Jonas, in contrast, aspires to understand life top-down: lower forms of life can be perceived through *privatio* of the features that higher forms exhibit. Jonas argues for continuity among life-forms. Knowing the highest form, human life, human beings are able to understand lower life-forms through an analogical inference based on extrapolated *privatio*.[38] Here, and in many other ways, Jonas is explicitly anthropocentric, and even anthropomorphic,[39] for he perceives the prevalent non-anthropomorphism to lead to absurdities:

> So radically has anthropomorphism been banned from the concept of nature that even man must cease to be conceived anthropomorphically if he is just an accident of that nature.[40]

Seen from another angle, Jonas's project of philosophical biology is to extend the Heideggerian hermeneutic of existence from humans to all living beings.[41] In this sense Jonas is less anthropocentric than philosophers who grant consciousness and personality only to human beings.

> Put at its briefest, this volume [*The Phenomenon of Life*] offers an 'existential' interpretation of biological facts. Contemporary existentialism, obsessed with man alone, is in the habit of claiming as his unique privilege and predicament much of what is rooted in organic existence as such: in so doing, it withholds from the organic world the insights to be learned from awareness of self. On its part, scientific biology, by its rules confined to physical, outward facts, must ignore the dimension of inwardness that belongs to life: in so doing, it sub-

> merges the distinction of 'animate' and 'inanimate.' A new reading of the biological record may recover the inner dimension—that which we know best—for the understanding of things organic, and so reclaim for the psycho-physical unity of life that place in the theoretical scheme which it had lost through the material and mental since Descartes.[42]

In the spirit of perceiving all life as analogous to human, of which one has first-person experience, and expanding the existential analysis, Jonas analyzes the dialectics of life. Common to all life is a reality that is inexplicable within dualistic metaphysics: *metabolism.*[43] Metabolism is the most basic phenomenon that separates animate beings from inanimate ones. While inorganic entities roughly preserve their identity over time, A=A remains A=A, organic beings through metabolism incorporate non-being into their existence, in at least a few respects even in the most simple life-forms, and therefore the organic identity appears less fixed than the inorganic. Moreover, an organic being has perpetually to achieve its identity. Non-being is involved in living beings, first because life is mortal; to live means to die eventually. Because of this fact, all life struggles to stay alive: an organic being constantly says "Yes" to life, to being, which means constantly saying "No" to death, to non-being. A second way in which non-being constitutes an essential part of organic, metabolizing life is inherent in metabolism itself: To metabolize is to have nutrition, consume matter, make use of it, and excrete. Metabolizing beings ingest matter and transform it. In this manner they incorporate non-being into their being.[44] This constant exchange and transformation of matter is the *sine qua non* of organic life:

> It is never materially the same, and yet it persists as this identical self by the very fact that it does not remain the same matter. If ever any two of its 'slices of time' do become identical with each other, then it has ceased to live: it is dead.[45]

As the above citation shows, Jonas refers to a living being as a "self." From this, it follows that he can speak of the relation of the self to the world and, further, that the self can go beyond itself toward the world, meaning *transcendence* is always present in organic life.[46] Moreover, the relationship of the self to the world implies that an organic being rests on the division between *exteriority* and *interiority*; and further, according to Jonas, interiority of the self is an *inwardness*, from which the organic being strives, and says "Yes" to its own being. Because Jonas understands life within this framework, his argument in support of *freedom* in all life is no surprise.[47] Jonas obviously perceives degrees to which different organic beings exhibit these characteristics. One of his main points, however, is that continuity always exists among life-forms and that not even in the case of human beings does an ontological gap or a categorical difference exist between species.

Jonas classifies organic life roughly into the same divisions as Aristotle. The first class includes metabolizing plant life, the second comprises animal life characterized by abilities of perception, motion, and feeling, and the third constitutes human life.[48]

The ascent in the hierarchy of organic life-forms, however, is no mere success story. As life in general carries death with it, higher life-forms encounter greater and greater risks. Jonas's concept is *needful freedom*: With increased freedom, inwardness, and transcendence come bigger needs and riskier means of finding satisfaction. Greater freedom and awareness always result from a precarious gamble,[49] but the gamble and its results are glorious:

> Measured in terms of mere biological security, the advantages of animal as opposed to plant life are highly questionable and in every case dearly bought. But survival alone is an inadequate measure for the evaluation of life. If it had been only a question of ensuring the duration, life ought not to have begun at all. It is by its very nature precarious and transitory, an adventure in mortality, and none of its possible forms is assured of duration as an inorganic being can be. Not duration as such but 'duration of what?' is the question here; in other words, such means of survival as perception and feeling are never to be judged only as means but also as qualities of the life to be preserved and therefore as aspects of the purpose of survival.[50]

Because it strives to maintain itself and push toward extensive freedom, despite the greater precariousness this entails, life bears inherent purposefulness. This purposefulness in organic life is the very link with which Jonas aims to ground ethics in ontology.

> However that may be (and the idea is frankly speculative), only an ethics which is grounded in the breadth of being, not merely in the singularity or oddness of man, can have significance in the scheme of things. It has it, if man has such significance; and whether he has it we must learn from the interpretation of reality as a whole, at least from an interpretation of life as a whole.[51]

Being and Ought

Jonas's theory of life manages to tackle the dualism part of dualistic nihilism: if life at least roughly is what Jonas argues life to be, it cannot possibly be divided into the compartments of *res cogitans* and *res extensa*, or their equivalents. The morality part of nihilism, however, still needs to be addressed. In tackling the problem of the arbitrariness of values, Jonas's theory must encounter two modern dogmas. These dogmas have so widely been held valid that at times they have been taken as hallmarks of progress in philosophy. The two dogmas are, first, the conviction that there is no metaphysical truth, that the task of philosophy is to undo whatever is left of traditional metaphysics, and second, that no "ought" can be derived from "is," the strict separation of fact and value—"Hume's Law."[52]

The dogmas are, according to Jonas, interrelated: Hume's Law is one instance of the more general denial of metaphysics. Jonas's simplest point concerning the dogmas is that both of them in themselves are metaphysical assertions. More particularly, they rely on the modern scientific, neutral notion of being. This is not the only possible way to conceive being, and therefore this approach itself is in need of metaphysical argument for its backing. If no argument is provided, one can conclude that modernity denies the possibility of every metaphysics other than its own. It is, in other words, argued simultaneously that no metaphysical truth exists and that it is a metaphysical truth that being is devoid of good. This is asymmetrical, to say the least.[53]

Furthermore, the same interrelatedness applies to ethics: all ethical theories contain tacit metaphysics embedded within them. For instance, moral theories that find the primary good in immanent pleasure indirectly suggest some form of immanentist metaphysics, perhaps materialism. Hence, it would be honest to open these metaphysical questions for discussion. Yet it is true that in metaphysics no final truths exist; sometimes various, even contradictory positions are under serious consideration simultaneously.[54] This is, however, exactly the inherent character of metaphysical pursuit:

> No physicist of today, for example, can rescue 'phlogiston' from history's graveyard. But Plato and Aristotle, Stoics and Epicureans, Hume and Kant, Hegel and Nietzsche offer continual subjects for debate for modern philosophers and can still find adherents among them. Whereas there can no longer be an alchemist or astrologer whom we take seriously, we can still take seriously an Aristotelian or Hegelian. In philosophy we cannot have a binding consensus about what is true and what is false. We cannot even desire one; it would spell the death of philosophy.[55]

In the spirit of traditional philosophy, Jonas grants that critique of previous metaphysics is possible and legitimate, but denial of metaphysics itself is not.[56] Jonas, in fact, considers metaphysics precisely the correct ground for finding the basis for ethics. He admits that religion (e.g., "the order of creation") can serve as a foundation for ethics. But religion works in the domain of faith, not of reason. Therefore, according to Jonas, it cannot be "on command" or provide strong and general enough arguments.[57]

Jonas practices, no doubt, bold metaphysical speculation. Yet the word "metaphysics" in his use has a less dramatic meaning as well. Often it appears to equal worldview (*Weltanschauung*), rational elaboration on both professed stances concerning the world picture and the not-so-consciously explicated aspects of the inevitable ideological frameworks that underlie human practices—everyday practices and the practice of science.

Jonas places his metaphysical-ethical argument concerning being and ought on various levels. His most fundamental discussion addresses the famous question posed by Gottfried Wilhelm Leibniz (1646–1716): Why is there something at all rather than nothing?[58] Jonas points out that the more relevant reading of the question would be: Why *ought* there be something at all rather than nothing? His reasoning behind this move has three parts. First of all, the latter question has more meaning, since it has a very practical application for humanity: it is in human hands whether being turns into nothingness. This point can be taken on at least two grounds. On the one hand, it reflects the novel and unique situation of humankind in the 20th and 21st centuries. On the other hand, one could understand it as a demarcation debate between ethics and ontology; Jonas might be arguing for the primacy of ethics. Second, Jonas argues that as a sole why-question, for example, for the cause of being, the question is idle and its answers beyond critical reflection.[59] Third, Jonas remarks that the traditional religious answers are not causal but emphasize the divine willed being, since it is *good.*[60] Thus, the notion of creation includes the idea that being exists not arbitrarily or due to a divine whimsy,[61] but because it ought to exist. Jonas, however, is not satisfied with these answers as mere revelatory truths, nor with the idea that being is good because God willed its existence. Jonas demands that this tradition be made meaningful through demonstrations based on being within the world, that being is good and therefore—if there must be a recourse to the divine—God saw it ought to be.[62]

This centers attention on the question: How can the preeminence of being over nothingness be shown? And the answer must be found in the world, available to anyone. In *The Imperative of Responsibility* this leads Jonas to explore ends and their status in reality. Jonas analyzes four examples: the hammer, the court of law, walking, and the digestive organ.

In the case of the hammer, its end exists before it. The hammer is made to realize an end that underlies its existence. Moreover, the end and purpose of the hammer exist outside the hammer: in its maker's mind and in his purposes of using the tool.[63] "[T]he concept here underlies the object, not object the concept as is the case with class concepts abstracted from, and thus subsequent to, things already existing."[64]

The court of law, a human institution, presents quite a different example of purposefulness from the hammer. First, in this case the ends are not visible in the physical appearance of the court. A mere physical description of court work cannot reveal its meaning. Material tools are obviously present in the court as well as in most social institutions; they, however, can only hint at the purpose of the legal action. Second, the ends are immanent in a manner that needs to be maintained by those who practice law; otherwise, they cease to exist. From this it follows, third, that the tools and buildings of the court cannot outlive its purpose. A thousand-year-old hammer, by contrast, is still a hammer even if its maker and users have passed away.[65]

The hammer and the court, nonetheless, have much in common. Both are artifacts and therefore their ends are human creations. Their origin and use also point out that they are not ends in themselves but refer to other ends outside themselves. In both cases, "the end in itself," says Jonas, appears to be "the human subjects whose ends they subserve."[66]

Walking exemplifies a voluntary organic function. In the case of walking, humans are liable to perceive legs as locomotive organs. *Organon* literally means "tool": the legs are transportation tools, and the human subject is understood as possessing the freedom to set goals for the use of these instruments. A person walks from place A to place B in order to be present in B. The presence is a condition for yet another end, so the chain of ends unfolds. How about walking animals? Are they understood similarly? On the one hand, it seems they could be. A cat stalks a mouse in order to attack it, in order to kill it, in order to eat, in order to quench its hunger. On the other hand, we would like to resort to concepts such as stimulus, response, instinctive action, and behavioral pattern in the case of the cat. In this manner the analysis of ends in walking leads Jonas to the problems of free will, inwardness, and the ontology of animal life. Although he does not manage to solve these problems fully, Jonas concludes that

> the realm of voluntary bodily movement in man and animal (exemplified in 'walking') is a locus of real determination by purposes and goals, which are objectively executed by the same subjects that subjectively entertain them. Implicit in this statement is the recognition that *efficacy of ends is not tied to rationality*, reflection, and free choice—that is, to men. *The efficacy is tied*, however, as far as our argument has gone, *to 'consciousness'* in some sense, *to subjectivity and to 'volition.'*[67]

In other words, Jonas shows that ends cannot be found in narrowly understood *res cogitans*, thinking substance, but they need to be looked for in life, or something that much resembles the Aristotelian soul. The example of the digestive organ takes Jonas a step further down in the ladder of organic life. In his analysis, the representational subjectivity and consciousness dissolve when he looks carefully at the most primal organic function. Diffuse appetitive striving, however, is still found even in the simplest life-form. In regard to this striving

> it is meaningful, and not just a metaphor borrowed from our subjectivity, to speak of the immanent, if entirely unconscious and involuntary, purpose of digestion and its apparatus in the totality of the living body, and to speak of life as the end purpose of just this body.[68]

The essential output of these explorations comprises the findings of *The Phenomenon of Life* reviewed previously: all life holds ends and purposes. Consequently, Jonas states:

> Thus, our showing up to now that Nature harbors values because it harbors ends and is thus anything but value-free has not yet answered the question of whether we are at pleasure or duty bound to join in her 'value decisions': whether, to put it paradoxically, the values undeniably entertained by and for herself are indeed valuable (even whether *having* values as such is valuable!)—in which case alone assenting to them would be a duty. . . . But showing the immanence of purpose in nature made asking this question now possible; and it will turn out that, with the gaining of this premise, the decisive battle for ethical theory has already been won.[69]

Although Jonas takes the refutation of the neutral concept of nature as a crucial victory, questions still remain unanswered. Most centrally, Jonas needs to show how the purpose-setting and value-positing that take place in nature are related to the Good in itself. For it is no news that agents strive for valued things: even a murderer, in some perverted sense, values the goal she or he aims to realize through a homicide. We would not, however, say that this valuation—at least without qualifications—captures something that is truly good. What then proves that values inherent in nature are not false? Jonas replies as follows:

> We can regard the mere *capacity* to *have* any purposes at all as a good-in-itself, of which we grasp with intuitive certainty [*von dem intuitiv gewiss ist*] that it is infinitely superior to any purposelessness of being. I am not certain whether this is an analytic or synthetic statement, but whatever of self-evidence it possesses, there is plainly no going back behind it for something more basic to underpin it. One can only oppose it with the doctrine of nirvana, which denies the value of having purpose, but then again affirms the value of liberation from it, making this its purpose. Since indifference is clearly not possible here (what is denied becomes a negative value), at least he who does not embrace the paradox of the purpose-denying purpose must concur in the proposition that purpose as such is its own accreditation within being, and must postulate this as an *ontological axiom*.[70]

Although my approach to this inquiry is first to provide a charitable description and then to pass into the critical considerations, I have to admit that Jonas's move is both ambiguous and question-begging. Instead of an argumentative *salto mortale* from worldly being to the binding good, we get recourse to intuitive certainty. I perceive here no real bridging of mere valuation act—which could be one of a hungry beast or a cunning murderer—and Good that obliges the (human) moral subject because of its being truly other and good independently of the valuations of the subjects. Moreover, given Jonas's focus on the primal forms of life, an apparently malicious virus—say HIV—might present a challenging borderline case for the ontologico-ethical intuition.

After expressing my first doubt, I can attempt a charitable account. A substantial, more empirical reading that is not looking for a strict philosophical demonstration would be that "the capacity to have purposes" is the simplest, most unconditional good in life—nay, in being. If the purposes cease, being is abolished. Jonasian being always includes good; good and being cannot be separated. This is the answer to the nihilist's badgering. Nihilism—science based and existentialist—rests on a false notion of being. If we bracket the hard-nosed analytic attitude, stand back, and take a wider look, it appears comprehensible and even compelling that human beings, who find themselves to be moral agents and observe all living beings striving for survival, ought to guarantee the living conditions of life. In this way, being issues an ought to humans: "*Being, in the testimony it gives of itself, informs us not only about what it is but also about what we owe to it.*"[71] When the same is looked at from the viewpoint of ethics, a discovery, radical in the modern conditions, is made: "a principle of ethics which is ultimately grounded neither in the authority of the self nor the needs of the community, but in an objective assignment by the nature of things."[72]

Lawrence Vogel has drawn a parallel between Heidegger's analysis of the call of conscience (*Ruf des Gewissens*) and Jonas's way to conceive the ought issued by being. Heidegger's *Dasein*, in the silent call of conscience, faces nothingness (*Nichts*) and is led to appropriate her or his ownmost possibilities in an authentic, or proper (*eigentlich*), resolve (*Entschlossenheit*).[73] According to Vogel, Jonas's call is not silent and empty but articulate: it has moral substance, because it emerges from the plenitude of being. This call, furthermore, works against the Kantian presuppositions of modern moral thinking, as in Jonas's reading being-nature-life turns out be a heteronomous, authoritative source of morality for the human being. Vogel formulates this in the following way: "The goodness of Being, reality, or nature opens up a '*genuine present*' because it gives us a future worth caring for."[74] And Jonas concludes:

> Its [the object of responsibility's call] highest duty, therefore, is to see that responsibility itself is not stifled, whether from its source within or from its constraints without. In the light of such self-transcending width, it becomes apparent that responsibility as such is nothing else but the moral complement to the ontological constitution of our *temporality*.[75]

Thereby Jonas perceives his project as thinking through the necessary ethical dimensions of the modern emphasis on the temporality of human existence. Our ontology—from Hegel, Nietzsche, and Darwin onward—is an ontology of time, not of eternity—as Plato's and still Spinoza's was. The object of *Eros*, or *amor intellectualis Dei*, is not in need of responsible care: it is complete and secured. Responsibility is needed amongst things that are temporal and therefore fragile and imperfect.

A more concrete way to elaborate on the manner in which a human being becomes the subject of the responsibility inherent in life is to look at archetypal relationships within which responsibility is encountered. A concrete example of such a relation can bridge the chasm between facts and values; Jonas writes as follows:

> [W]hen asked for a single instance (one is enough to break the ontological dogma) where the coincidence of 'is' and 'ought' occurs, we can point at the most familiar sight: the newborn, whose mere breathing uncontradictably addresses an ought to the world around, namely, to take care of him. Look and you know.[76]

Thus, Jonas presents the parent-child relationship as the archetype of responsibility. He admits that a reductionist scientific account of an infant does not encounter in an infant any ought but just conglomeration of cells that are reduced to molecules, to atoms, to microparticles, and so on. To be able to see the baby, however, the human being must make use of her visual faculty,[77] and as the subject of this organic function she will be faced with the ought. Hence, "look and you know."[78]

TOWARD UPDATED ETHICS

Thus far I have followed the key movements with which Jonas prepares a way for a new ethic appropriate for the predicament of humanity in modernity. This new morality is the ethics of responsibility that attempts to respond to the problems of humankind faced with its own fearsome powers and the alarming reactions of the planetary ecosystem to human technological activities. Responsibility, according to Jonas, is a correlate of power:

> Only in man is power emancipated from the whole through knowledge and arbitrary will and only in man can it become fatal to him and to itself. His capacity is his fate, and increasingly becomes the general fate. . . .[79]
>
> [T]hat which binds will and obligation together in the first place, *power*, is precisely that which today moves responsibility into the center of morality.[80]

The new ethic that Jonas visions concerns more public policy than personal morality. For thinking about responsibility in this domain, another archetype of the responsibility relationship needs to be looked at, next to the parent and the child: that of the statesman's bond to her or his subjects. Just as a parent owes the possibility of being a responsible parent to her or his parents, statesmanship takes place on a succession:

> The general principle here is that any total responsibility, with all its particular tasks and all its single actions, is always responsible also for preserving, beyond its own termination, the *possibility* of responsible action in the future—that is, for preserving its own preconditions.[81]

Thus responsibility is both transitive and self-referring. This is central in Jonas's updated ethic, the core of which is the imperative of responsibility. The imperative has many formulations, such as the following:

> 'Act so that the effects of your action are compatible with the permanence of genuine human life'; or expressed negatively: 'Act so that the effects of your action are not destructive of the future possibility of such a life'; or simply: 'Do not compromise the conditions [necessary] for the indefinite continuation of humanity on earth'; or, again turned positive: 'In your present choices, include the future wholeness of Man among the objects of your will.'[82]

For the reasonable and sensitive, these imperatives may appear intuitively acceptable. It would not, for instance, be easy to look one's child or grandchild in the eyes if one had serious doubts concerning the ethos articulated in the imperatives. Earlier parts of this chapter have, however, exhibited that the intellectual justification for such imperatives is far from clear, or even absent, if the mainstream of modern culture is taken for granted. Since Jonas's suggestions for the *Grundlagen* reform have already been introduced, I shall elaborate on the conceptual clarifications of the imperative.

A central point concerning the context of the imperative of responsibility Jonas terms the "heuristics of fear." It refers to a rule of interpretation according to which, in the prognosis for the future, prophecies of doom need to be given priority over the prophecies of bliss. Modified for moral principles, it is the idea that avoidance of the greatest *malum* ought to precede the pursuit of the greatest *bonum*.[83]

These principles counter both classical Western ethical-political thinking, in which the pursuit of *bonum* has been crucial, and the modern thinking, which has transposed eschatology into the secular realm. In philosophy the most important manifestations of the secular eschatology have been the Marxist utopianism and the Nietzschean reflections concerning the future of man. Marxism conceived socialism both as the historically determined next stage of humanity and as the duty of those conscious of the state of history to realize it. Nietzsche perceived humankind at the threshold as the post-Christian era opened in Europe. This situation could produce either the nihilistic, comfort-seeking last man or *Übermensch*, or overman—a state of being beyond human, some kind of new heroic spontaneity free of the guilt, repression, and *ressentiment* distinctive of the era of Western metaphysics and Christianity.[84] Common to these very different modern ways of thinking is the idea of *bonum* as a goal of historical succession.[85]

Jonas analyzes and criticizes the Marxist utopianism especially in its form presented by Ernst Bloch (1885–1977) in *Das Prinzip Hoffnung* (1959), to the name of which *Das Prinzip Verantwortung*—the German title of Jonas's book—alludes. Tackling Marxist utopianism is philosophically crucial for Jonas, since Marxist thinking had appropriated Baconian technoscientific optimism and integrated it into its own ethical-political progressivism.

> A critique of utopia has become necessary with the seeming possibility of its realization. For the first time in the annals of man, thanks to the powers of technology, the dream appears to be capable of turning to a task, and Marxism has seized on this novel chance to give its political gospel eschatological exaltation and pragmatic credibility at the same time.[86]

These passages in Jonas's 1984 book, however, have become somewhat dated since the collapse of the Soviet Union and the Eastern Bloc. Moreover, Nietzsche's hopes for the overman have never been widely accepted in the mainstream of modern thought, although his analysis of the last man has inspired cultural critics.[87] Utopias, nonetheless, are far from obsolete. In both implicit and explicit forms they adhere closely to the theme of this inquiry: technology. A new kind of humanity is currently sought, not in a reformed societal order or transformation of existence, but with the aid of various technologies. Frontiers such as transhumanism or cyborgs are the exotic manifestations of this pursuit, critiques of posthumanism its portrait in the hands of the concerned. Behind the slogans of the search for the cure for cancer, gene therapies, and stem cell research, for instance, lie utopian perspectives of health and long life.[88] I suggest that narratives on utopias still form the background of many technological projects, at least in an implicit manner.

Against all kinds of utopianism, Jonas argues that ethics of responsibility is "[b]orn out of danger, its first urging is necessarily an ethics of preservation and prevention, not of progress and perfection."[89] Heuristics of fear, since Jonas's book's publication, has become a part of ordinary parlance under the name of precautionary principle. The concept of responsibility, in fact, has made its way into the vocabulary of public policy as well. Consider "corporate responsibility," "social responsibility," and "environmental responsibility." The common uses of these terms, however, need to be kept separated from Jonas's own thinking.

The ethics of responsibility is not ethics of ascent but one of survival: heuristics of fear relates to Jonas's requirement that the generation living at the turn of the millennium should push the unquestioned quest for abstract affluence aside, and concentrate on and grant primacy to the very tangible questions of survival.[90] This demand seems to be in good balance with Jonas's ontology that emphasizes the material, organic side of human beings.

Jonas begins his theory of responsibility with a distinction between formal and substantive responsibility. Formal responsibility is mere attribution of causality in a deed: the person who effected the deed is responsible for it. In this sense responsibility means accountability and comes close to legal liability. It, for instance, is crucial in theories of punishment: the degree of punishability is dependent on the degree of being responsible for the deed. Yet responsibility in this sense is not Jonas's main concern:

> the affirmation of formal responsibility is not a sufficient principle of morality and its pure formality cannot supply the affective (emotive) element for ethical theory.[91]

Concerning substantive responsibility, Jonas emphasizes that adequate ethics for the technological world must have both a rational, or objective, and an emotive, or motivational, side. The objective side refers, on the one hand, to the objective value in which ethics, according to Jonas, should be grounded. On the other hand, it refers to objective knowledge, science foremostly, that needs to be taken into account in ethics in a technological era. As I pointed out previously, Jonas disapproves of a dualism that would place ethics as subjective, consciousness related, or strictly human in a sphere opposite objective science. Therefore, Jonas's account of organic life is not merely philosophical but attempts to stay in dialogue with the modern biosciences. In the context of the imperative of responsibility, this line of thought means that ethics of responsibility should be considered in tandem with scientific study of the future, futurology.[92]

The specific character of the *feeling* of responsibility—the emotive, motivational element—is due to the character of the object of responsibility. The object of most philosophically conceived moral feelings, Plato's *Eros*, Spinoza's *amor intellectualis Dei*, or Aristotelian *hedone*, relates to some kind of supreme good. It is difficult to think about responsibility in conjunction with it. This is due to differences in the objects:

> By contrast the object of *responsibility* is emphatically the perishable *qua* perishable. Yet in spite of this condition which it shares with myself, it is more unsharably an 'other' to me than any of the transcendent objects of classical

> ethics; 'other' not as surpassing better, but as nothing-but-itself in its own right, and without *this* otherness being meant to be bridged by a qualitative assimilation on my part or on its part. Precisely this otherness takes possession of my responsibility, and no appropriation is intended here.[93]

The germ of this feeling Jonas finds in the archetypal responsible relationships, parents feeling responsible for their children and statesmen feeling responsible for their subjects and the future of the state. These archetypal feelings, nevertheless, will transform in the face of dire circumstances created by technological advance. Perishability has a novel and unique form in these conditions. Therefore, the birth of this new kind of feeling depends on the dialogue with knowledge disclosing the future, read in light of the heuristics of fear: the optimistic tones must not override the prophesy of doom.[94]

In sum, being aware of tasks of responsibility with the aid of scientific knowledge forms the rational substance of substantive responsibility, and the feeling that can grow from the experience of parenthood and statesmanship forms the emotive, motivational substance of it.

In chapter 2 I briefly noted that in environmental stances, various degrees of anthropocentrism exist. What is Jonas's place in the continuum between anthropocentrism and holism? What is it that he thinks the present generations are responsible for? From his premises Jonas draws the following explicit conclusion:

> Since in him [man] the principle of purposiveness has reached its highest and most dangerous peak through the freedom to set himself ends and power to carry them out, he himself becomes, in the name of that principle, the first object of his obligation, which we expressed in our 'first imperative': not to ruin (as he well can do) what nature has achieved in him by the way of his using it.[95]

Thus Jonas's argument proceeds to characterize human life as the epitome of a purposeful life that ascends toward higher forms of freedom.[96] And therefore, the duty to guarantee the possibilities for genuinely human life lies at the core of the imperative of responsibility as we already perceived in its formulations. To capture the notion that responsibility particularly concerns securing the humaneness of human life, Jonas postulates "the idea of Man":

> With this imperative we are, strictly speaking, not responsible to the future human individuals but to the *idea* of Man, which is such that it demands the presence of its embodiment in the world. It is, in other words, an ontological idea, which does not, however, as the 'ontological proof' alleges concerning the concept of God, guarantee the existence of its subject already with the essence—far from it!—but says that such a presence *ought* to be and to be watched over, thus making it a duty to us who can endanger it. It is this

> ontological imperative emanating from the idea of Man, that stands behind the prohibition of a *va-banque* gamble with mankind. Only the idea of Man, by telling us why there should be men, tells us also *how* they should be.[97]

Thus, responsibility consists of the present generation reflecting on its own interests against the idea of Man. A generation may aim to do something, and it may be in its own best interests, but if it does not pass the test of preserving genuinely human life—*id est*: it is not in harmony with the idea of Man—then the imperative of responsibility urges us to renounce such action. Human cloning, for instance, if it became possible and began to be practiced, would distinctly alter human life, and hence such a practice should be assessed against the idea of Man.[98]

Then the next big question emerges: What kind of life is genuinely human? What is the content of the idea of Man? This question requires a recourse to anthropology. Jonas's anthropology appears to have certain developments, but in *The Imperative of Responsibility* he depicts humans as life-forms that are characteristically capable of being responsible:

> To be *de facto* responsible in some respect for someone at some time (whether acknowledging it or not) belongs as inseparably to the being of man as his *a priori* capacity for it—as inseparably indeed as his being a speaking creature—and is therefore to be included in his definition, if one is interested in this dubious pursuit [that Jonas is up to in *IR*].[99]

His argument therefore forms a kind of "benign circle": The imperative of responsibility binds the present generation to the obligation of securing the continuation of responsible, genuinely human life: "Only the idea of Man, by telling us why there should be men, tells us also how they should be." This intergenerational responsibility exists as a structure and potentiality in the archetypes of the parent and the statesman. Both have the responsibility to secure the continuity of their own roles. The parent aims at bringing up the children so that they will one day be able to be parents, and the statesman attempts to guarantee the existence of statesmanship, for instance, through taking care of the political institutions. Although Jonas's anthropology exhibits traits of anthropological essentialism, it is fairly open-ended, since future generations are free to rethink the human good in many ways, and only subduing the *homo responsibilis* would be questionable.[100]

Jonas's anthropology has other facets in addition to the responsibility central in *The Imperative of Responsibility.* Based on his analyses of dualism and organic life, one easily understands Jonas's emphasis that human beings share many characteristics with other organic beings. As organic life-forms, humans cannot be cleanly analyzed into the compartments of mind and matter. This ontological point also lies beneath Jonas's controversial 1970 critique of brain death:

> I see lurking behind the proposed definition of death, apart from its obvious pragmatic motivation, a curious revenant of the old soul-body dualism. Its new apparition is the dualism of brain and body. In certain analogy to the former it holds that the true human person rests in (or represented by) the brain, of which the rest of the body is a mere subservient tool. Thus, when the brain dies, it is as when the soul departed: what is left are 'mortal remains.'[101]

Some extent of dualism, however, needs to be preserved in order to provide an adequate account of how human beings differ from other animals. A totally monist materialist account of being loses the distinctive character of being human. Against this background, a crucial feature in Jonas's anthropology is the emphasis on mortality. In the texts "Burden and Blessing of Mortality" (1991) and "Tool, Image, and Grave: On What Is beyond the Animal in Man" (1985) Jonas sets mortality in the center of his anthropology.[102] He argues that at the core of humanity lies the human being's awareness of her limited time on earth, the inevitability of death. Consciousness of mortality, according to him, is a more important aspect of humanity than tool-making (*homo faber*) or image-making (*homo pictor*). Furthermore, mortality and the awareness of it, although they burden the human life, work as important incentives in existence, as well. This is obviously an ancient philosophical theme, and a Heideggerian one indeed.[103] Yet it also characteristically Jonasian: only beings who are aware of their own perishability can truly be responsible for other perishable beings. Jonas's point, in short, is that eliminating the reality of mortality technologically—for instance, via prolonging the life span radically or with strong palliatives—results in life that lacks something, which would be genuinely human.

> Knowing his mortality, he must live as a human being with a self-image that is by no means self-evident but is the tentative result of questioning and speculation. This speculation necessarily expands from the individual ego to the whole of existence, in which the ego finds itself situated. Thus, metaphysics arises from graves.[104]

Jonas points out that the responsibility relationship most often—maybe always—exists between parties that are in unequal positions, such as parents and their children. Therefore, the relationship is asymmetrical and nonreciprocal.[105]

* * *

The difficulties addressed in the previous sections are not only intellectual in nature: The equation in which one attempts to balance ecological sustainability, democracy,[106] a liberal understanding of rights, consumption levels common in developed nations, a growing population, economic growth in developing countries, a global economy framed around competition, and more, embodies a concrete problem of unequaled magnitude. To tackle seriously the task of solving such a problem would require considerable renunciation worldwide, and particularly in developed nations. Human history has seen no instance of such voluntary renunciation. Jonas's ethics of responsibility essentially endeavors to set these unparagoned requirements into a meaningful and justifiable intellectual framework.

CRITICAL REFLECTIONS

Next, I make a few critical remarks on Jonas's thinking. I aim to do this in a fruitful spirit, to advance the themes he started, not to refute his project. I shall begin by reflecting on the character and necessity of Jonas's speculative thinking, then move on to the remarks on the imperative of responsibility and ethical naturalism as its justification. The fourth section addresses Jonas's anthropology, the fifth the problem of evil in his thinking, and the sixth the concept of technology.

Digesting Metaphysics

In its own era, Jonas's metaphysical thinking was extraordinary. Whatever disagreements and total lack of understanding of different approaches philosophers might have had in the 20th century, they agreed that metaphysical systems and speculations belonged to the past. Against this background, the obvious reactions to Jonas's metaphysics could be the following. A philosopher from the analytic tradition might ask: In what way would this theory be falsifiable? If it is in no way falsifiable, what is the point of discussing it? A purist phenomenologist would point out that Jonas goes far beyond *die Sache selbst* given for consciousness. A biologist would be interested to know whether the latest research in biosciences, in fact, supports Jonas's ideas. Such criticisms are understandable. Jonas obviously was well aware of these viewpoints, and the motive behind his theories was the need to find a ground for ethics so that it would be saved from being merely arbitrary. Lawrence Vogel has, however, asked: What is the worth of such grounding, if the grounding itself unavoidably appears arbitrary? Vogel gives credit to Jonas's analyses of nihilism and dualism that enable us "to see things differently" but is skeptical of his constructions.[107] I find Vogel's critique valid.

Moreover, it is hard to understand how one would find a proper sense of responsibility within the context of organic life or the idea of humanity. These are terribly abstract notions. If Jonas's metaphysics is found to be both arbitrary and lacking potency in motivating moral action, one might want to reopen the question concerning religion as the background story of ethics. Philosophically, religion may be even more arbitrary than Jonas's metaphysics, but religiously understood responsibility is most often more concrete and motivating than philosophical theories. Moreover, religious traditions work against subjective idiosyncrasies.[108]

Although Jonas's metaphysics as metaphysics raises doubts, it has other kinds of merits. As one strong motive for Jonas's ontological thinking I perceive his desire to reconcile the modern natural scientific worldview and his deeper needs for comprehensibility. On the one hand, modern natural science alone, especially in the forms that the strongly pro-science reductionist approaches put forth, is not able to provide a view of the world that would make a human being "feel at home in the universe" in the manner that the ancient views of cosmos were able to do. On the other hand, various forms of anti-science thinking[109] that discredit science as a provider of the true picture of reality put contemporary people in an odd position: humankind is totally dependent on technoscience, but somehow science and technology are projects gone astray or without any firm basis. Thus, discrediting science appears to lead to estrangement as much—or even more than—rigorous reliance on science. Jonas's merit is that he manages to avoid both extremes and suggest a synthetic position, the price of which is the uniqueness of his metaphysics.

The Character of the Imperative

The previous question relates to the criticism of the imperative of responsibility: it appears to be formal and very general. This raises the question of what, in fact, is the real and distinctive content of the imperative. Lawrence Vogel asks: How can the imperative of responsibility be applied to public policy? Richard Bernstein inquires: To what extent does the responsibility articulated mean responsibility for the idea of humanity or for concrete others? Jonas emphasizes the idea of Man, but responsibility for only an idea would be an awkward notion. There should be some relation to real people.[110] The question needs to be asked the other way around as well: Who is the bearer of Jonas's imperative? It is mainly for public policy, but what institutions should it bind and how? These are not substantial critiques but points that will need further clarification if Jonas's approach is to be taken seriously.

Karl-Otto Apel addressed Jonasian themes in substantial depth in his essay "The Problem of a Macroethic of Responsibility to the Future in the Crisis of Technological Civilization" (1987). I consider two points in Apel's essay relevant in the context of this inquiry. First, Apel raises questions concerning Jonas's contention that the ethics of responsibility for contemporary humanity demands primarily—and maybe merely—guaranteeing the existence of humankind, irrespective of its progress. Apel suggests that Jonas's stance would allow suspect measures for securing the existence of humankind. If continued human life on earth is the sole goal of the ethics, this could entail, for instance, letting some portions of the human population die. This Social Darwinist solution might secure the living conditions for those who stayed alive. The moral doubts concerning such measures, however, would require recourse to some principles of moral progress, argues Apel. Therefore, Jonas's denial of utopianism and progress appears untenable, as it may lead to morally suspect conclusions.[111] Apel asks:

> [One can] ask oneself whether the existence and dignity of the human being can be saved at all by *merely preserving the now existing state of affairs*. More precisely, is the nature of man and of his long since technologically and socioculturally transformed environment not so constituted that without a *regulative idea of technological and social progress* it cannot be *preserved*?[112]

Apel's point is relevant. Yet it loses most of its dramatic edge when one reads both Apel and Jonas more carefully. What Apel refers to as ideas of progress in his own thinking turn out to be Kantian regulative principles of reason, which Apel grasps as a transcendental condition for a community of rational communication. Thus he is speaking about progress in a more formal and minimal sense than Jonas, in whose writing utopianism and progress come close to each other. Jonas, moreover, raises the idea of Man to a normative idea that should provide the frame for the future of humanity. Apel's counterexample of the Social Darwinist approach might, however, pass the test set up by Jonas, as its main target is the preservation of humanity. What Jonas's inference— "only the idea of Man, by telling us why there should be men, tells us also *how* they should be"—in this context would mean is not fully clear. In sum, Apel's substantial point I perceive to be the following: Is life without justice a worthwhile goal?

Second, the common interest for both Apel and Jonas is the foundational set of problems in ethics. They both perceive it to be severe and to a large extent resultant of technoscientific advance. Both, furthermore, think this issue should be addressed and some constructions for its solution found. Their solutions are simply different. Apel does not agree that the foundation for ethics needs to be sought in "objective" nature. The key word in Apel's approach to grounding ethics is "universal": the foundation for ethics is implicit in the conditions for communicative rationality:

> [T]he *ought*, in the sense stipulated by the ethics of discourse, can be derived from the deciphered 'fact of reason,' that is, from the fact of always already having recognized the duties belonging to a member of the real as well as of the anticipated ideal community of communication.[113]

In general, Jonas's analysis of nihilism, the groundlessness of ethics, appears to underestimate Kantian ethics and contract theories as foundational avenues. Even if no ground for ethics can be found in nature, other solid grounds may exist. Human autonomy and dignity, universalism implicit in reason, and the contract character of existing human sociability may provide a footing for ethical thinking. A Rawlsian would probably remark to Jonas that the problem of nihilism is an artificial construction, of which the strengths of John Rawls's (1921–2002) approach provide evidence. Jonas thus appears to go too far in discrediting the non-naturalist ways of ethical thinking. Furthermore, in addition to these rationalistic and mostly deontological approaches to ethics, more feeling-based, for instance Humean, ways of ethical thinking are still available for consideration.[114]

Ethics and Being Considered

As earlier described, Jonas makes a long argument from the problems involved in the idea of value-natural being to justifying the imperative of responsibility. Long arguments nearly always contain weak points. Two steps in Jonas's inference are especially vulnerable to critique. First, despite the insightfulness of his reflections, Jonas is not able to make a tight, positive argument that being always involves good. Even if organic being is inevitably purposeful, it is not an undoubtable proof that being *in general* involves good. Another weak point is the step from the purposefulness in organic nature to human duty. With Jonas's reading of instructions for nature, this step makes sense. In addition to benign purposefulness, there are many kinds of phenomena in nature, however. If we take natural catastrophes, predators, viruses, and cancer cells into account, reading nature as a moral guide becomes more difficult than in Jonas's interpretation.

* * *

In addition to the structure of Jonas's argument, two other points are worth reflection. First, in what sense is the teleology *in re* in living nature? In what sense is the teleology in the human being's way of framing phenomena? In other words, what is the relationship of ontology and epistemology here? These are standard philosophical questions that Jonas might answer by pointing out that they arise from the modern worldview that is the target of his critique. Phenomenology and Heideggerian ontology, moreover, claim not to be trapped in the representationalist thinking underlying my distinctions of epistemology and ontology. To an extent, I would find answers along these lines satisfactory.

Second, I am not sure how well Jonas manages to integrate the viewpoints of biology and phenomenological analysis. The integration undoubtedly is a worthwhile goal and Jonas's avenue of "descending anthropocentrism" a fruitful approach toward the goal. It just appears that "a leap" persists between the human consciousness, which is the center of the phenomenological approach, and both the organic life, which is to be understood through Jonas's analysis, and methodological naturalism, which is the framework, within which biological knowledge is acquired. In other words, phenomenology *per definitionem* studies the consciousness from within; it is transcendental work. We cannot, however, have this "from within" perspective on life or consciousness in other kinds of living beings. Moreover, whether it is necessary concerning the substance or not, the modern biosciences have developed exclusively based on the *objectifying* and *experimental* description of life-forms. How can phenomenological analysis that is either "subjective" or beyond subject-object dichotomy make use of natural scientific knowledge that emphatically relies on the notion of object? In slightly other terms, how far "down" could Jonas extrapolate the human firsthand experience among the life-forms? Based on the analysis of life in us, it appears relevant to think about life in other mammals. Life in reptiles, nevertheless, would be more of a challenge. The level of primal life, amoebas for instance, in which metabolism happens in its simplest form is central to Jonas's argument, and it is not very clear how well the extrapolation from the human realm "reaches" that level. Life in the cell scale is just abstraction for the human mind: it is accessible to us rather through experiments, research instruments, and theories than through experience and observation.

Jonas's Anthropology

The trouble of combining Jonas's insistence on both anthropocentrism and the human ability to understand organic life other than human points toward Jonas's anthropology. Jonas's anthropocentrism is understandable in the sense that as humans we have to start our inquiry into organic life from our own *anthropos*-perspective. Jonas attempts to find a balance between two extreme, unsatisfactory but logically consistent positions. At one end of the continuum are the monistic, biocentric, deterministic stances: mere life within which a human being has no special metaphysical or moral status. At the other end are the dualistic, anthropocentric, voluntaristic stances: the mainstream of humanistic and Judeo-Christian tradition, which has not had much concern about life-forms other than human.[115] Given his task, Jonas's middle way obviously appears sound and wise, but strictly from the viewpoint of methodical thinking it has a slight flavor of an ad hoc solution.

In addition to the overall character of Jonas's anthropology, there is one curious omission in its content. In his argument for the inevitability of ethics, Jonas makes use of the case of a newborn. He does not, however, address the question of the imaginary skeptic in the beginning of this chapter: Why is the baby there in the first place? Even if we always felt an obligation to take care of an infant when we came across one, we could still ask whether it is a good idea to produce more babies in the world. Jonas does not really address this question. His argument that being is better than non-being—if the argument is accepted as valid—provides a general background answer as to why it is better to have more human beings than to refrain from reproduction. Many details of this idea are, however, missing. Most importantly, Jonas, as Leon Kass has pointed out, does not address sexuality in his anthropology. This is an omission, given the references to parent-child relationship as an archetype of responsibility and the description of responsibility against the contrast of the Platonic *Eros*.[116]

In very practical terms, Jonas does not give reasons why the newborn is there and why she ought to be there. One explanation would be that human beings cannot help but procreate because of their sexual desire. Another could be that humans want to have children. From a slightly different angle, one could propose that humans want to see their life or their family continue. An overall naturalistic viewpoint would explain that all organic beings multiply and therefore it is according to nature that humans propagate as well. In earlier times, not much real choice as to whether to have children or not existed. The technological age has, however, changed the conditions for procreation.[117] For the first time in history, humans have safe means of contraception. Most importantly, the too-large number of human beings on the globe is a threat to the biosphere and to humans themselves. Further, as the chances of future humans surviving weakens, we may reach a point where it may not be moral any longer to have people born into such bad conditions. Finally, the desire to see one's own children grow is at least to an extent selfish. One can well argue that such a desire is in need of more legitimation.

The Problem of Evil

I find it paradoxical that Jonas, who had faced great evils in his lifetime, was still so optimistic in his philosophy. Jonas suggests that purposeful willing in life is good. A pessimist may still ask: On what grounds can Jonas assume, as he seems to, that willing and striving in life are always good? Could not willing itself be destructive and cause suffering? Especially after witnessing how well-intentioned technoscientific efforts have created an apocalyptic ecological crisis, to suggest that the volition behind this action has been corrupt would not be unreasonable. In the Western philosophical tradition, Arthur Schopenhauer's (1788–1860) philosophy would provide a fitting, all-encompassing pessimistic interpretation of the world.[118] In Schopenhauer's thinking, all that exists is inherently manifestation of will. In their will to live, human beings and other animals seek happiness but end up causing suffering to themselves and other living beings. Therefore, Schopenhauer argues the soundest goal for life is the denial of will, for instance through compassionate action. For Schopenhauer the project seeking technoscientific progress is just another manifestation of the blind, cruel will. Moreover, in his reading, values are purely functions of willing. In brief, where Jonas sees an organism reaching out toward good, Schopenhauer sees blind will positing values.

Within Christianity, the doctrine of original sin exhibits a form of "partial pessimism": the human will can never be fully good and innocent.[119] Practical implications of such interpretations of willing would be limiting or renouncing the expressions of the will. This obviously is manifested both individually and collectively in the ascetic practices of many cultures, especially those in Asia.[120] In fact, Heidegger's practical remedies for the predicament of the technological era resemble these pessimistic solutions. He proposed *Gelassenheit*, detachment or "releasement," as the wisest attitude toward technology, since on the one hand, wholesale acceptance of technoscience disables meditative thinking in favor of calculative thinking, which results in losing the proper relation to Being. Demonizing technology, on the other hand, becomes as entangled with technology as enthusiasm, only through negation. Therefore, the best way, according to Heidegger, to live with technology is with an attitude of detachment.[121] Because this study does not address Jonas's theology, I formulated the above considerations as a question of how Jonas's metaphysics manages the problem of evil. Jonas does not pay attention to natural catastrophes or cruelty in nature. I perceive that the problem remains open in Jonas's philosophy. His theology is yet another story.

Finally, I cannot help wondering how relevant Jonas's narrative of Gnosticism is. Jonas's elaboration on the predicament of technological modernity through an analogy with Gnosticism comes close to the interpretations of Christianity as the scapegoat for environmental problems.[122] One problem with Jonas's—admittedly insightful—interpretation is that neither Gnosticism nor Heidegger represent the mainstream of the Western culture—during their time or after it. Another question is whether the light provided by the parallel is correct. Undoubtedly, Heidegger and Gnosticism appear to have plenty in common. But mainstream modern culture seems to be very far from the otherworldly Gnostic attitudes. If we, the modern people, perceive the world as bad, why do we so enthusiastically work and mold the world? Why are we so eager to gather property? Why are safety, sex, and health all-important to us if we find the world, and the body, worthless?

In any case, I find "the theology of technology," interpreting religious and quasi-religious stances related to technology, meaningful. Technology and religion easily remain as invisible others for each other, although they seldom or never exist totally separate from each other.[123] Georg Henrik von Wright (1916–2003), a philosopher more well-known as a logician, found a kind of "magical" attitude toward reality beneath modern technoscientific pursuits. He pointed out that most earlier forms of philosophy, religion, and science—in the manner of Jonas's typological Stoicism—had left the world as it is and concentrated on improving the human being through the cultivation of virtue and submitting oneself to the laws of reality. Modern technoscience, however, manipulates reality—through scientific experiments with technical apparatuses—in order to attain knowledge, in order to manipulate reality—with technical devices created in conjunction with systematic research, in order to advance human happiness. This process essentially involves no notion of reality as a norm for the human behavior in it. Thus, reality is no longer any authority, but human desire is, and the attitude of submission to the world has turned into a stance of manipulating the world. This stance, von Wright hints, has much in common with the prescientific magic.[124] The typological magician molds the world to please his own will; he is not interested in the world in itself nor in improving his own character. Von Wright argues that technoscience shares this volitional stance with magic. I think von Wright's story of manipulative ethos as magical is more relevant than Jonas's analogy with Gnosticism.[125]

If von Wright's story is relevant and valid enough, one viable remedy for the culture in the manipulative frenzy could be the above-mentioned Heidegger's *Gelassenheit*—be it Gnostic or not.

Jonas's Notion of Technology

Although *The Imperative of Responsibility* proposes an *Ethics for the Technological Age* in its title, Jonas spends very little time analyzing technical phenomena or elaborating on technology itself. What the human action in technology is, and what kind of entities technological artifacts and systems are, remain most of the time opaque. Jonas mainly treats the consequences of technological development and pays little attention to technology itself. I find this an omission. Yet such an omission is understandable, as this study has come to witness the elusiveness of technology as a theme of philosophical reflection. Still, I think a deeper analysis of the phenomenon of technology is required of an ethics of technology. On this point, Jonas's work could have benefitted from a closer dialogue with Heidegger, Ellul, and Mumford, who elaborated on the character of technology. Today, those finding the Jonasian approach worth developing would do well to update the notion of technology with both a philosophy of technology and science and technology studies.

CONCLUSION

Most concluding remarks concerning Jonas have already been articulated. A couple of points, however, are still worth mentioning. First, let us return to the reflections of the imaginary skeptic that opened this chapter. How differently does one perceive such disillusioned thoughts after a brief journey into Hans Jonas's thinking? I think the soundest answer for the skeptic would be to ask him or her to inquire into the framework within which his or her questions are posed. According to Jonas's analysis, the skeptic clearly subscribes to the modern idea of value-neutral nature and therefore acknowledges no pressing obligation to take care of the living conditions of future generations. His or her critique, nonetheless, remains troubling. Jonas's thinking, to be honest and realistic, would not convert a person holding such a position. Yet his thinking is aimed at converting people. For Jonas it is not enough to accept an argument here and another there; his is not a piecemeal approach. His goal is to make people see things, the whole world, life, and humanity, in a new light. How often do people, in fact, let themselves "be converted"? How often do we address our fundamental metaphysical—the word used in Jonas's sense here—convictions and habits of thought? And when we do, how does a change in them happen? These questions point at the limits of philosophy. This study cannot give definite answers to them. Still, I think it has become evident that it is meaningful for the ethics of technology to plunge into metaphysical analysis. In the context of technology, ethics can ill afford to leave untouched such conceptions as the modern scientific idea of neutral nature. Metaphysical constructions, however, at least in Jonas's case, are problematic. Both analyses and constructions enable one to see things in a different light. This apparently is the main goal of analysis, and thus results are appropriate. Argumentation for large-scale metaphysical positions, however, appears at times arbitrary, at times slightly embarrassing. This in part is due to the present stage of our culture; this is not an era of building new systems of thought.

Second, concrete human life in the world of technology has so far been absent in this study. Although Jonas addresses the human condition with seriousness, his thinking remains in the large issues of public policy and in broad historical trajectories. Nevertheless, a smaller perspective is relevant in ethics as well: What is everyday life like in a technological world? What does the migration from the farm to the assembly line, from the factory to the office cubicle, from the constant work to the realm of entertainment mean to the subject of ethics? In the next chapter, I look at phenomena that are most often taken for granted, since they lie so close to the subject of the technological world that they present no exotic topic on which to philosophize.

NOTES

1. This fictitious critic of politically correct notions was inspired by Matti Häyry's column in "the city newspaper" *Metro Helsinki* (January 5, 2006). The above text does not, however, directly represent Häyry's views. Those interested in arguments in this direction may find Häyry's articles, e.g., "If You Must Make Babies, Then at Least Make the Best Babies You Can?" (*Human Fertility*, 2004) and "A Rational Cure For Prereproductive Stress Syndrome" (*Journal of Medical Ethics*, 2004) useful.

2. Hans Jonas, *Wissenschaft als persönliches Erlebnis* (1987), 14–17; "*Wissenschaft* as Personal Experience," *Hastings Center Report* (2002), 29–30; "Heidegger's Resoluteness and Resolve," *Martin Heidegger and National Socialism: Questions and Answers* (1990), 197–200; "Philosophy at the End of the Century," *Mortality and Morality* (1996), 44; *Memoirs* (2008), 39–72. Richard Wolin, *Heidegger's Children: Hannah Arendt, Karl Löwith, Hans Jonas, and Herbert Marcuse* (2001), 101–104; "Ethics after Auschwitz: Hans Jonas's Notion of Responsibility in a Technological Age," *The Legacy of Hans Jonas: Judaism and the Phenomenon of Life* (2008), 59–72. Richard Bernstein, *Radical Evil: A Philosophical Interrogation* (2002), 163–165, 184–187. David Levy, *Hans Jonas: The Integrity of Thinking* (2002), 1–4, 24–28.

3. "My second reservation has to do with Heidegger's conduct in year 1933. Does that have to do anything with philosophy? In my opinion, yes. Since ancient times philosophy, unlike every other branch of learning, has been guided by the idea that its pursuit shapes not only the knowledge but also the conduct of its disciples, specifically in the service of the Good, which is after all the goal of knowledge. . . . The example of Socrates, which has served as a beacon for philosophy since its beginnings, has kept the belief in such ennobling force from being extinguished" (Jonas, "Philosophy at the End of the Century," 49). See also "Heidegger's Resoluteness and Resolve," 200–203.

4. In a radio interview in 1987 Jonas ("Heidegger's Resoluteness and Resolve," 200–201) describes his first reflections on the problem as follows: "In 1933, when he gave that infamous rectorial address, justifiably called treacherous in a philosophical sense and actually deeply shameful for philosophy, I was simply appalled and spoke with friends about it and said: 'That from Heidegger, the most important thinker of our time.' Whereupon I heard the reply: 'Why are you so surprised? It was hidden in there. Somehow it could already be inferred from his way of thinking.' That was when I realized, for the first time, certain traits in Heidegger's thinking and I hit myself on the forehead and said: 'Yes, I missed something there before.'" See also Wolin, *Heidegger's Children*, 131–133.

5. Jonas's most straightforward treatment of Heidegger is found in "Heidegger and Theology" (*Phenomenon of Life*, hereafter *PL*, 235–261), given as an opening address in a theological conference at Drew University in 1964. Jonas's address was preceded by a number of theses that Heidegger had sent for the conference. Jonas, *Memoirs*, 187–193. See also Wolin, *Heidegger's Children*, 101–104.

6. Jonas acknowledged that his experiences in the wars affected his philosophy. In fact, his wife sent books on biology to the front, as Jonas had turned his philosophical attention from ancient religion to understanding organic life. Jonas, *Wissenschaft als persönliches Erlebnis*, 20–21; "*Wissenschaft* as Personal Experience," 31–32; *Memoirs*, 110–148. Wolin, *Heidegger's Children*, 104–108.

7. Jonas, *Memoirs*, 149–186. Bernstein, *Radical Evil*, 163–165, 184–187.

8. Jonas, *Wissenschaft als persönliches Erlebnis*. Christian Schütze, "The Political and Intellectual Influence of Hans Jonas," *Hastings Center Report* (1995). Wolin, *Heidegger's Children*, 107–108; "Ethics after Auschwitz," 1–15. Levy, *Hans Jonas*, 7–8. In an introduction to a collection of philosophical essays that contains texts on all his fields of interest, Jonas (*Philosophical Essays: From Ancient Creed to Technological Man*, 1974, xi–xviii) explains the contingent historical background of his philosophical path and presents some intellectual links between the fields. The latter will be elaborated on, to some extent, in what follows in this chapter.

9. See, e.g., Jonas, *PL*, 262–281; *Mortality and Mortality*, hereafter *MM,* 115–143, 165–197; Wolin, *Heidegger's Children*, 129–131; Bernstein, *Radical Evil*, 194–199; Lawrence Vogel, "Hans Jonas's Exodus: From German Existentialism to Post-Holocaust Theology," *Mortality and Morality: A Search for the Good after Auschwitz* (1996), 19–36; Levy, *Hans Jonas*, 95–131.

10. This distinction in the deficiencies of ethics corresponds to my categories of adequacy questions and foundation questions in the second chapter. Jonas addresses my first category of "challenges" as well. Of this his critical analyses of possible human cloning (Jonas, "Biological Engineering—A Preview," *Philosophical Essays*, 141–167) and the concept of brain death ("Against the Stream: Comments of Definition and Redefinition of Death," *Philosophical Essays*, 132–140) serve as good examples.

11. Jonas, *Philosophical Essays*, hereafter *PE*, 3–20. The prospect of creating a new kind of humanity in tandem with modern science and technology, nevertheless, was already present in Bacon's and Descartes's visions. See, e.g., *Novum Organum*, I.129 and part six of *Discourse on Method*. "For even the mind is so dependent on the temperament and on the arrangement of the organs of the body that, if it is possible to find some means that generally renders men more wise and more capable than they have been up to now, I believe that we must seek for it in medicine" (René Descartes, *Discourse*, translated by Richard Kennington, 2007, 49).

12. Jonas, *Imperative of Responsibility*, hereafter *IR*, 22.

13. Jonas, *PE*, 45–80; *MM*, 47–49; *PL*, 213–217.

14. See, esp., the second, fifth, and sixth *Meditations* and parts IV–VI in *Discourse on Method*. Jonas's reading of Descartes is not the most charitable. More refined interpretations are possible. For instance, Sara Heinämaa (1999) analyzes the ways in which emotions (*passiones*) work as a bridge between *res cogitans* and *res extensa* in human beings.

15. Jonas, *PL*, 13–17, 37.

16. Hans Jonas, *Gnostic Religion: The Message of the Alien God and the Beginnings of Christianity* (1958/2001), 320–340/211–234 (the essay was published both in *Gnostic Religion* and *The Phenomenon of Life*). An earlier version of the essay was published in 1952 under the title "Gnosticism and Modern Nihilism" in *Social Research* 19 (1952, No. 1), 430–452.

17. What follows is of course an ideal-typical description of the Gnostic existential-metaphysical stance, since no uniform Gnosticism ever existed. In his books on Gnosticism Jonas gives a more nuanced account of the movement. For example, Michael Waldstein in his "Hans Jonas' Construct 'Gnosticism': Analysis and Critique" (*Journal of Early Christian Studies*, 2000), offers a contemporary critique of Jonas's scholarly studies on Gnosticism. Antti Marjanen ("What Is Gnosticism? From the Pastorals to Rudolph," *Was There a Gnostic Religion?* 2005, 40–45) and Kurt Rudolph ("Hans Jonas and Research on Gnosticism from a Contemporary Perspective," *The Legacy of Hans Jonas: Judaism and the Phenomenon of Life*, 2008) draw more balanced assessments of Jonas's contribution to the Gnostic studies. On the role of Gnosticism in Jonas's thought, see Kalman P. Bland, "Reflections on the Place of Gnosticism and Ethics in the Thought of Hans Jonas," *The Legacy of Hans Jonas: Judaism and the Phenomenon of Life* (2008). Studies on Gnosticism were revived by the discovery of a Gnostic library in Nag Hammadi, Egypt, in 1945. It, however, took some time for the texts to become accessible for research, and the field is still developing. *Was There a Gnostic Religion?* (edited by Antti Marjanen, 2005) offers a spectrum of views concerning the present understanding of uniformity in Gnostic movements.

18. Jonas, *Gnostic Religion*, 42–96; "Gnosticism," *The Encyclopedia of Philosophy* (1967); *PL*, 221–228.

19. Jonas, "Gnosticism," 341.

20. *Being and Time*, esp. §§ 27, 35–38, 44, 53, 60. Graham Harman (*Heidegger Explained: From Phenomenon to Thing*, 2007, 173) scrutinizes the distinction as follows: "The more humans interpret themselves on the basis of entities, the more inauthentic they are. The more humans interpret themselves in terms of their own inherent potentiality for being, the more authentic they are."

21. Jonas, *PL*, 213–217, 224–232.

22. Jonas, "Philosophy at the End of the Century: Retrospect and Prospect," *MM,* 44–49; *PL*, 227–234.

23. Heidegger, *Being and Time*, §§ 29, 31, 38.
24. Ibid., § 40.
25. Obviously, the question still remains open as to what are the proportions of Heidegger's analysis as a description of human existence in the modern world and its influence on the interpretation of the human condition.
26. Jonas (*PL*, 244–249) addresses *Schicksal* in the lecture "Heidegger and Theology," in which he warns the Protestant theologians of the inherent "paganism" of Heidegger's thought.
27. Heidegger, *Being and Time*, §§ 46–53. The summary of death's role goes as follows: "We may now summarize our characterization of authentic Being-towards-death as we have projected it existentially: *anticipation reveals to Dasein its lostness in the they-self, and brings it face to face with the possibility of being itself, primarily unsupported by concernful solicitude, but of being itself, rather, in an impassioned freedom towards death—a freedom which has been released from the Illusions of the "they," and which is factical, certain of itself, and anxious*" (*Being and Time*, 1962, 311, emphases in the original). "Die Chrakteristik des existential entworfenen eigentlichen Seins zum Tode lässt sich dergestalt zusammenfassen: *Das Vorlaufen enthüllt dem Dasein Verlorenheit in das Man-selbst und bringt es vor die Möglichkeit, auf die besorgende Fürsorge primär ungestütz, es selbst zu sein, selbst aber in der leidenschaftlichen, von den Illusionen des Man gelösten, faktischen, ihrer selbst gewissen und sich ängstenden Freiheit zum Tode*" (*Sein und Zeit*, 2001, 266).
28. Jonas, *PL*, 215. In this analysis Jonas comes, in fact, quite close to Heidegger's understanding of nihilism in "Word of Nietzsche: 'God Is Dead'" (*The Question Concerning Technology and Other Essays*, 1977, 53–112; *Gesamtausgabe, Band 5: Holzwege*, 2003, 210–247).
29. Jonas, *PL*, 47, 181, 211–234, 282–284. In my view, Jean-Paul Sartre's existentialism shares the same problem, although Sartre is somewhat eager to cultivate some kind of existential moralism, which *mutadis mutantis* resembles Kantian ethics. It does not, however, appear cogent in the context of his existentialism, where logically the only meaningful imperative would be "choose yourself, don't follow conventions" (Sartre, "Existentialism Is a Humanism," *Existentialism and Humanism*, 1965; Jonas, *PL*, 225–226.) Further, as Heidegger frequently points out (*Being and Time*, § 38), there cannot be a moral imperative even for commitment; from where would it come? Authenticity (*Eigentlichkeit*) is only a possible mode of existence toward which *Dasein* may strive out of its everyday inauthenticity (*Uneigentlichkeit*). (*Being and Time*, esp. §§ 27, 34–38, 53–60).
30. In addition to cosmos and *logos*, *phusis* is a pregnant notion here. Jonas evidently operates in a metaphysical space opened by Heidegger, for instance, in *Introduction to Metaphysics* (2000). Jonas's approach differs from Heidegger's most importantly in its constructive aims. Heidegger's "metaphysics" is indirect and deconstructive, a project of *Destruktion* of Western metaphysics. Jonas, however, aims at constructing new metaphysical ways of thinking. Below, Jonas's theory of organic life will exhibit a way in which Jonas attempted to reawaken a classical, mostly Aristotelian conception of *phusis* through reinterpretation of modern biology.
31. Jonas, *PL*, 213–217, 232–234. Vogel, "Hans Jonas's Diagnosis of Nihilism: The Case of Heidegger," *International Journal of Philosophical Studies* (1995a), 56–61. Levy, *Hans Jonas*, 11–34. Christian Wiese, *The Life and Thought of Hans Jonas: Jewish Dimensions* (2007), 98.
32. Jonas, *PL*, 196.
33. Jonas, *IR*, 22–24; "Toward an Ontological Grounding of an Ethics for the Future," *MM*, 101–103; *PL*, 282–284.
34. Jonas, *PL*, 53. Vogel, "Hans Jonas's Exodus," 12. In an interview (Jonas and Harvey Scodel, "An Interview with Professor Hans Jonas," *Social Research*, 2003), Jonas describes Husserlian phenomenology and the phenomenon of hunger as follows: "For instance, I will give you an example. With Husserlian means, with Husserlian phenomenology, you may give a wonderful account of what you experience with the feeling of hunger, and perhaps also a very good description of what you experience with the stilling of hunger, with the satisfaction of your need for food . . . of what conscious phenomena are involved there. But this account is unable to raise the question, How much does the body need? Man must eat. How much? In proportion to the size of his body, and in proportion to the size of the environment. That in a great degree determines the human condition and is of fundamental importance. And in phe-

nomenology you haven't the descriptive categories to deal with that question. There, you have to enter into a quite humble relationship, a knowing relationship with what science tells you. Why is there a recurrent feeling of hunger? I may phenomenologically give a very good account of what the experience of taste is, and of quenching of thirst, and so on, but it is an entirely different thing to know how much liquid my body needs and when lack of liquid becomes a danger."

35. Jonas, "Evolution and Freedom: On the Continuity among Life-Forms," in *MM*, 59–74; *PL*, 7–26, 64–92.

36. Jonas, *PL*, 11.

37. Jonas, "Technology as a Subject for Ethics," *Social Research* (1982), 894–895; "Toward an Ontological Grounding of an Ethics for the Future," *MM*, 101–103; *PL*, 282–284. Leon Kass, "Appreciating *The Phenomenon of Life*," *Hastings Center Report* (1995), 14–17.

38. Jonas, "Evolution and Freedom," *MM*, 67; *PL*, 22–26, 33–37. Susanna Lindberg, "Hans Jonas's Theory of Life in the Face of Responsibility," *Phänomenologische Forschungen* (2005).

39. Jonas, *PL*, 10, 33–37, 91, 117, 122, 278, 282.

40. Ibid., 233.

41. In the sections §10 and §12 of *Being and Time* Heidegger touches upon the points from which Jonas departs. "Life, in its own right, is a kind of Being; but essentially it is accessible only in Dasein. The ontology of life is accomplished by way of a privative Interpretation; it determines what must be the case if there can be anything like mere-aliveness [Nur-noch-leben]. Life is not a mere Being-present-at-hand, nor is it Dasein. In turn, Dasein is never to be defined ontologically by regarding it as life (in an ontologically indefinite manner) plus something else" (*Being and Time*, 1962, 75). "Leben ist eine einige Seinsart, aber wesehaft nur zugänglich im Dasein. Die Ontogie des Lebens vollzieht sich auf dem Wege einer privativen Interpretation; sie bestimmt das, was sein muss, dass so etwas wie Nur-noch-leben sein kann. Leben ist weder pures Vorhandensein, noch aber auch Dasein. Das Dasein wiederum ist ontologisch nie so zu bestimmen. Dass man es ansetzt als Leben – (ontologisch unbestimmt) und als überdies noch etwas anderes" (*Sein und Zeit*, 2001, 50). Heidegger discusses at length the problem of animal being in *The Fundamental Concepts of Metaphysics* (1995).

42. Jonas, *PL*, xxiii.

43. Jonas ("An Interview with Professor Hans Jonas") aims at overcoming dualism with a form of monism that is neither materialist nor idealist/spiritualist: "To give a monistic account, but not in favor of one of the two sides, not an option for A or B, but, if possible, an understanding of their both belonging to a full inventory of reality. That is, for a doctrine of being, for a theory of being, account must be taken of the phenomenon of life, however rare it may be in the universe." In *De anima* (III.12, 434a 22–26) Aristotle characterizes the nutritive soul (*threptikê psykhê*) as metabolism that all the living beings have: "Every living thing, then, must have the nutritive soul, and in fact has a soul from its birth until its death; for what has been born must have growth, a highest point of development, and decay, and these things are impossible without food. The nutritive faculty must then exist in all things which grow and decay." "Τὴν μὲν οὖν θρεπτικὴν ψυχὴν ἀνάγκη πᾶν ἔχειν ὅτι περ ἂν ζῇ καὶ ψυχὴν ἔχῃ, ἀπὸ γενέσεως καὶ μέχρι φθορᾶς· ἀνάγκη γὰρ τὸ γενόμενον αὔξησιν ἔχειν καὶ ἀκμὴν καὶ φθίσιν, ταῦτα δ' ἄνευ τροφῆς ἀδύνατον· ἀνάγκη ἄρα ἐνεῖναι τὴν θρεπτικὴν δύναμιν ἐν πᾶσι τοῖς φυομένοις καὶ φθίνουσιν." Many threads of influence can be read out of Jonas's account. Aristotle is probably the most obvious influence; others include Friedrich von Schelling, G. W. F. Hegel, Charles Darwin, and Alfred North Whitehead.

44. Jonas, "Evolution and Freedom," *MM*, 60–66; *PL*, 3–6, 75–87.

45. Jonas, *MM*, 64.

46. "By 'transcendence' of life we mean its entertaining a horizon, or horizons, beyond its point-identity" (Jonas, *PL*, 85).

47. Jonas, *MM*, 63–70; *PL*, 80–92.

48. Jonas, *MM*, 70–73; *PL*, 99–107. Lindberg, "Hans Jonas's Theory of Life in the Face of Responsibility."

49. Jonas, *MM*, 66, 73–74; *PL*, 80, 86.

50. Jonas, *MM*, 73.

51. Jonas, *PL*, 284. Here one can notice how Jonas transposes the meaning of being (*to on*) toward nature (*physis*) and life (*psyche, zoe, bios*). This move again draws on Heidegger's work (*Introduction to Metaphysics*, 14–19, 64, 106, 119), but Jonas's emphasis on organic life, purposefulness, and the good separates him from Heidegger (*Introduction to Metaphysics*, 210–214: Heidegger on the good).

52. Jonas, *IR*, 44–45; *MM*, 193–194. William LaFleur, "Infants, Paternalism, and Bioethics: Japan's Grasp of Jonas's Insistence on Intergenerational Responsibility," *The Legacy of Hans Jonas: Judaism and the Phenomenon of Life* (2008), 465–466. David Hume in his *Treatise* (1958, 469–470, chapter 3.1.1) originated the law that an inference from plain "is" to "ought" is impossible. Hume writes as follows: "I cannot forbear adding to these reasonings an observation, which may, perhaps, be found of some importance. In every system of morality, which I have hitherto met with, I have always remark'd, that the author proceeds for some time in the ordinary way of reasoning, and establishes the being of a God, or makes observations concerning human affairs; when of a sudden I am surpris'd to find, that instead of the usual copulations of propositions, *is*, and *is not*, I meet with no proposition that is not connected with an *ought*, or an *ought not*. This change is imperceptible; but is, however, of the last consequence. For as this *ought*, or *ought not*, expresses some new relation of affirmation, 'tis necessary that it shou'd be observ'd and explain'd; and at the same time that a reason should be given, for what seems altogether inconceivable, how this new relation can be a deduction from others, which is entirely different from it. But as authors do not commonly use this precaution, I shall presume to recommend it to the readers; and am persuaded, that this small attention wou'd subvert all the vulgar systems of morality, and let us see that the distinction of vice and virtue is not founded merely on the relations of objects, nor is perceiv'd by reason." Hume's point has created a massive dispute concerning its interpretation. Jonas does not address Hume directly. Therefore, it is not relevant to go into Hume's exegesis here.

53. "Thus, with the very assumption of such a concept of being, the rigid separation of 'is' and 'ought' reflects in itself already a definite metaphysics, which can only boast the critical (Occamistic) advantage that it makes the most parsimonious assumption of being (therewith, of course, also the most meager for the purpose of explaining the phenomena, hence at the price of the impoverishment of the latter). But if the dogma that no path leads from 'is' to 'ought' is, by reason of its ontological presupposition, a metaphysical proposition, then it falls under the interdict of the first and more fundamental dogma: that there is no metaphysical truth. This proposition has its own presupposition to which its validity is tied. . . . [T]he denial of metaphysical truth presuppose[s] a definite concept of knowledge, for which it is indeed true: 'scientific' knowledge is not to be had about metaphysical objects—once again a tautological conclusion since science is just concerned with physical objects" (Jonas, *IR*, 44).

54. Jonas, *IR*, 44–45.

55. Jonas, *MM*, 41–42.

56. Ibid., 193.

57. Jonas, *PE*, 169–170; *PL*, 282–284. Vogel, "Hans Jonas's Exodus," 19–21. Jonas called theology "the luxury of reason": it does not work on the same plane of understanding and binding rationality as philosophy.

58. Leibniz addressed the issue in his 1697 essay "On the Ultimate Origination of Things" (*Philosophical Essays*, 1989, 149–155). Although Jonas does not make explicit references or critiques, his treatment of the question seems to be in dialogue with Heidegger's examinations of the same question—for instance, in *Introduction to Metaphysics*.

59. Jonas's relationship with metaphysics appears to be ambivalent indeed: at one point he argues for its inevitability and at another he wants to curb it.

60. Jonas, *IR*, 47–48. Compare: Genesis 1:1, 1:31; Plato, *Timaeus*, 29d–31a.

61. This line of thought is in harmony with Jonas's critique of Gnosticism: in some Gnostic myths the world is created by a mistake. Antti Marjanen, "What Is Gnosticism? From the Pastorals to Rudolph," 36. Jonas, "Gnosticism."

62. Jonas, *IR*, 47–48.

63. Ibid., 44–45.

64. Ibid., 44.

65. Ibid., 53–56.

66. Ibid., 55–56.
67. Ibid., 65 (emphasis added).
68. Ibid., 75.
69. Jonas, *IR*, 78 (emphasis in the original). Vogel ("Hans Jonas's Diagnosis of Nihilism," 63) elaborates on this central Jonasian theme as follows: "Once we recognize mechanistic materialism for what it is—a metaphysical prejudice, not a neutral description of the physical world—we are free to interpret the biological facts existentially and to appreciate the reality of value independent of us, for all organisms are 'ends-in-themselves' who value whatever contributes to their existence and welfare."
70. Jonas, *IR*, 80.
71. Jonas, "Toward an Ontological Grounding," *MM*, 101 (emphasis in the original).
72. Jonas, *PL*, 283.
73. Heidegger, *Being and Time*, §§ 54–60.
74. Vogel, "Hans Jonas's Exodus," 14–15.
75. Jonas, *IR*, 107 (emphasis in the original).
76. Ibid., 131.
77. For Jonas's account on the phenomenology of sight see *PL*, 135–156. I do not think, however, that Jonas's "look and you know" is strictly dependent on sight. Without awkwardness one can think that blind people can "look" by making use of other senses.
78. Jonas, *IR*, 130–135. Despite its insightfulness, the example of the infant raises many questions. I first wonder to what extent it is in line with Jonas's "organistic" account. Is the ought issued due to the baby being an organic being or more her or his being a person? Jonas probably would answer that it is due to both: personhood and organic being cannot be separated without artificiality. The question still remains which one of these perspectives is more fundamental as the basis for ethics. Second, what is the difference of the ought one feels for one's own baby and someone else's? The crying of a strange baby often sounds merely annoying, while one's own child's crying indicates suffering. This question hints at the problems of nepotism and the distance of universal ethics from Jonas's example (see Wolin, *Heidegger's Children*, 121–122).
79. Jonas, *IR*, 129.
80. Ibid., 130.
81. Ibid., 118.
82. Ibid., 11.
83. Ibid., 26–27, 202–203.
84. Nietzsche, *Thus Spoke Zarathustra* (2006), 5–12, 93–95, 231–236. Yet as I indicated earlier, the possibility of a new kind of humanity was already entertained by Francis Bacon and René Descartes, the principal thinkers in the birth of modern technoscientific project.
85. Jonas, *IR*, 157–158, 178–204.
86. Ibid., 178.
87. Jonas touches only briefly on "Superman" and concludes as follows: "[T]here is not a hint in Nietzsche what can be done concretely to bring about the higher man, to enable and expedite his appearing, or even merely the probability of it (unless one takes occasional similes from breeding as such hints)" (*IR*, 158). Jonas's judgment of the current Western culture, however, is not very far from Nietzsche's feared last man. In an interview in 1990 ("An Interview with Professor Jonas"), he remarks as follows: "You see, the so-called good life, what's now in the West, in our entirely cheapened and degenerate value system is (said) to be the good life—this is a life of plenty, of a very vulgar kind that itself reflects a (vulgar) conception of the Good."
88. See, e.g., Ray Kurzweil, *The Age of Spiritual Machines: When Computers Exceed Human Intelligence* (1999); Bill Joy, "Why the Future Doesn't Need Us," *Wired* (2000); Francis Fukuyama, *Our Post-Human Future: Consequences of the Biotechnology Revolution* (2002); Langdon Winner, "Are Humans Obsolete?" *The Hedgehog Review* (2002); Jürgen Habermas, *The Future of Human Nature* (2003). Karl-Otto Apel ("The Problem of a Macroethic of Responsibility to the Future in the Crisis of Technological Civilization: An Attempt to Come to Terms with Hans Jonas's 'Principle of Responsibility,'" *Man and World*, 1987, 9) raises the point that an "involuntary" or "automatic" utopianism might exist within the background as-

sumptions of the advanced form of life. Jonas does not thematize this possibility. This is probably because he wants to avoid the Heideggerian "fatalism" in facing the problem of the technoscientific civilization. Nevertheless, I perceive the clarification of such a background idea an important task, although it would probably require more "deconstructive" thinking than Jonas's.

89. Jonas, *IR*, 139.
90. Ibid., 26–27.
91. Ibid., 92.
92. Ibid., 25–31.
93. Ibid., 87 (emphases in the original).
94. Ibid., 108–130.
95. Ibid., 129. The German version goes as follows: ". . . da in ihm das Prinzip der Zweckhaftigkeit durch die Freiheit, sich Zwecke zu setzen, und die Macht, sie auszuführen, seine höchste und zugleich selbstbedrohende Spitze erreich hat, so wird im Nahmen des Prinzips er sich selber zum ersten Gegenstand des Sollens, nämlich jenes erwähnten "ersten Gebots", nicht das ihm Erreichte, wie er *auch kann,* durch die Art seiner Nutzung zu verderben" (Jonas, *Prinzip Verantwortung*, 1979, 232).
96. In "Tool, Image, and Grave: On What Is beyond the Animal in Man" (1985), Jonas (*MM*, 75–86) discusses the characteristics of human life. He ends up emphasizing an understanding of finiteness in the form of mortality, whereas earlier in his career he gave more weight to *homo pictor*. See also "Burden and Blessing of Mortality," *MM*, 87–98.
97. Jonas, *IR*, 43.
98. Jonas wrote an influential critique of (possible) human cloning: "Biological Engineering—A Preview" (*PE*, 141–167). The crux of Jonas's argument was that it would be unethical to create clones, since as duplicates of a person who has already lived ahead of them the clones would lack the ignorance of their genetic potentials; seeing one or more outcomes of the potentialities would affect their freedom. Jonas appealed to 'the right to ignorance' (163): "The ethical command here entering the enlarged stage of our powers is: never violate the right to that ignorance which is a condition for the possibility of authentic action; or: *to respect the right of each human life to find its own way to be surprised to itself.*" See also Bernard G. Prusak, "Cloning and Corporeality," *The Legacy of Hans Jonas: Judaism and the Phenomenon of Life* (2008). For contemporary positions on cloning see John Harris, "Cloning," *A Companion to Applied Ethics* (2003).
99. Jonas, *IR*, 99.
100. Ibid., 38–44.
101. Jonas, *PE*, 139. The essay "Against the Stream: Comments on the Definition and Redefinition of Death" was written in 1970 and published in 1974 in *Philosophical Essays.*
102. Jonas, *MM*, 75–98.
103. The theme of mortality is obviously Jewish as well. Psalm 90 (verse 12), for example, asks, "Teach us to number our days aright, / that we may gain a heart of wisdom." In Aeschylus's *Prometheus Bound* (1991, 148–149, lines 249–256) Prometheus's account of his gifts to human beings combines forgetfulness of doom and the ability to learn crafts: "*Chorus*: Did you perhaps go further than you have told us? *Prometheus*: Yes, I stopped the mortals from foreseeing doom. *Chorus*: What cure did you discover for that sickness? *Prometheus*: Sowed in them blind hopes. *Chorus*: That was a great help you gave to men. *Prometheus*: Besides, I myself gave them fire. *Chorus*: Do now the creatures of the day own the bright-faced fire? *Prometheus*: Yes, and from it they shall learn many crafts."
104. Jonas, *MM*, 84.
105. Jonas, *IR*, 93–94.
106. Jonas perceived democracy as an ill-suited political order for encountering the ecological challenges of late modernity: "[I]n the preceding deliberations it was tacitly assumed that, in the coming severity of a politics of responsible abnegation, democracy (in which the interests of the day necessarily hold the stage) is at least temporarily unsuited, and our present day comparative weighing is, reluctantly, between different forms of 'tyranny'" (*Imperative of Responsibility*, 150–151).

107. Vogel, "Does Environmental Ethics Need a Metaphysical Grounding?"; "Hans Jonas's Exodus," 36–40.

108. This suggestion entails thinking about religion in pluralistic conditions, since one single religion obviously cannot serve as "the background story" for all the people on the globe. Detailed reflections on this topic reach beyond the scope of this study. Yet it is not implausible to think that people from different religions facing the same global-ecological condition could find the meaning and motivation from their religions to tackle the common crisis. Although plausible, this line of thought falls exactly in the dualistic track that Jonas attempted to avoid: "objective ecological situation" is interpreted in the light of "subjective religion."

109. Anti-science thinking can take off from several starting points. The phenomenological tradition since Husserl and Heidegger (e.g., "The Age of the World-Picture," *The Question Concerning Technology*, 115–154), has possessed viewpoints for a fundamental critique of modern natural science. Some extreme forms of social constructionist science studies have been able to make science look as irrational, taboo- and myth-driven a pursuit as any other human activity—religion, politics, or fashion (Harry Collins and Trevor Pinch, *The Golem: What You Should Know about Science*, 1998, 150–181). In addition to these academic viewpoints, different activist stances exist. They may be based on religious doctrines or environmentalism. Moreover, academic *skepsis* and activist pursuits may be combined, which can result in socially reconstructed science in accordance with ideological goals. Some branches of biotechnology (e.g., stem cell research), research on climate change, and debates on evolution, for instance, are frontiers on which science and worldviews mingle (on the political character of science, see Carl Mitcham and Robert Frodeman, "New Directions on the Philosophy of Science," *Philosophy Today*, 2004).

110. Vogel, "Hans Jonas's Exodus," 40. Bernstein, *Radical Evil*, 199–201.

111. Apel, "The Problem of a Macroethic of Responsibility," 1–8, 18–20.

112. Ibid., 7.

113. Ibid., 32 (emphasis in the original).

114. Vogel ("Hans Jonas's Exodus," 39–40) suggests the Humean alternative.

115. Wolin (*Heidegger's Children*, 121–122), for instance, worries that Jonas may undermine humaneness and the human good as he emphasizes mere life and the animal. For him Jonas's anthropology appears too organic, too little human. A deep ecologist might launch a critique from the other end of the continuum.

116. Kass, "Appreciating *The Phenomenon of Life*," 11–12. Kass argues that in omitting sexuality Jonas follows Heidegger: the human being is an alone, gender-neutral being facing the world. Moreover, the emphasis on death as the humanizing reality is Heideggerian. Kass would like to see more attention given to sexuality, love, and marriage.

117. These alterations in the human condition in general are the main theme of Jonas's *The Imperative of Responsibility*.

118. See esp. §§ 57–65 in *The World as Will and Representation*.

119. See, e.g., Romans 5:12–21, 7:7–25; Ephesians 3:1–3.

120. In contrast to the Western traditions, in the Asian traditions "pessimism" is more the mainstream than a marginal strand.

121. Heidegger, *Discourse on Thinking: A Translation of* Gelassenheit (1966); *Gelassenheit* (2000a); *The Question Concerning Technology*, 23–35.

122. See, e.g., Lynn White's *Machina ex Deo* (1968).

123. Bruno Latour (*We Have Never Been Modern*, 1993, 31–32, 39–43, 50–59, 78–79) has coined the term "purification" for the purpose of his analyses of modernity. Purification refers to separations of this type. In Latour's use purification refers to the modern attempts to observe society (politics, meaning, human doing) and nature (extensional, measurable, "objective" reality) separately.

124. G. H. Von Wright, *Vetenskapen och förnuftet. Ett försök till orientering* (1986), 38–54; *Ciencia y razón: una tentativa de orientación* (1995); *Tieto ja ymmärrys* (1999), 42–53. Topi Heikkerö, "The Fate of Western Civilization: G. H. von Wright's Reflections on Science, Technology and Global Society," *Bulletin of Science, Technology, and Society* (2004).

125. Von Wright is not alone with his ideas of a manipulative ethos. C. S. Lewis's small book *The Abolition of Man* (1943) presents very similar ideas. Moreover, Heidegger in "The Age of the World-Picture" (*Question Concerning Technology*, 115–154) comes close to these themes as well. A curious common denominator between Jonas and von Wright is Oswald Spengler's thinking (*The Decline of the West*, 1945, originally 1918–1922), since both appear to have been inspired by Spengler's characterizations of the Arabic culture as the source of the magical approach to reality. On Jonas's connection to Spengler see Jonas's *Memoirs*, 86–87.

Chapter Four

Albert Borgmann

Technology and the Good Life

The role of science in human life, in simple terms, is to make reality more comprehensible. For example, teaching a child that the necessary conditions for fire are burnable substance, oxygen, and sufficient temperature deciphers the mystery of fire to an extent. In addition to the step in comprehension, the simple concepts help the small human being to live with fire: to make it, to tend it, and to extinguish it. In this way a proto-scientific piece of knowledge can make one more at home in the world, both through understanding and ability to act. Some technologies seem to involve this aspect of scientific thinking to a great extent. For instance, the theory behind the combustion engine—four-stroke, two-stroke, or diesel—is quite straightforward, and knowing this theory is crucial in the maintenance of engines. In a way, simple physics, a motored vehicle, skills needed to take care of the vehicle, and using it—at least in principle—can all form an understandable whole. Many technologies, however, do not belong to this category. The personal computer, for example, offers a convenient operating system and inviting programs to the user. But underneath this surface complex software is at work, on many levels, and this programming rests on equally complex hardware. Although the user can know some general principles of binary numbers and electronics, the way the computer works stays, for the most part, hidden underneath the pleasant user interface.

Somewhat paradoxically, science seems to have stripped nature of most of its mysteries while, simultaneously, modern technology has created a mysterious human-made world of devices and networked infrastructures.[1] Just consider an international flight and all the processes needed to make it happen and how little we know of them. What kind of effect has this development had on human beings? They have found their way out of dangerous and unknown nature to a user-friendly and yet unknown human-made world. How does this affect them as human beings?

Albert Borgmann addresses this question mainly by asking how technology has affected the human possibilities for the good life. This question belongs to "the first facet" of the ethics and technology relationship that was framed in chapter two. The question, however, penetrates the other two facets as well: the notion of the good life most probably needs reconsideration in technologized conditions (second facet), and "life ethical" thinking must be intellectually grounded in order to achieve tenability—beyond mere wishful thinking or cultural cosmetics—in philosophical clarification on the theme (third facet). Borgmann's roots are in Southern Germany:

> I was born and raised in Freiburg, in the shadows of a Gothic cathedral, the Black Forest, and the university where Husserl and Heidegger had been teaching. My parents were part of Freiburg Catholic intelligentsia and acquainted with Bernhard Welte, Karl Rahner, and Max Müller—liberal Catholic thinkers who had come under the spell of Heidegger.[2]

He studied both in Germany, where he was born in 1937, and in the United States, where he earned an MA in German literature from the University of Illinois before returning to the University of Munich to earn his PhD in philosophy. The latter he completed under Max Müller. Heidegger's thinking on technology has influenced Borgmann strongly. One does not, however, notice this in Borgmann's writing, which is clear and transparent in the manner of the analytical philosopher's. Rather, Heidegger's influence is seen in the content and ethos of Borgmann's work. He himself comments that his style and method have been influenced by John Rawls, whose *Theory of Justice* (1971) made a distinctive impression on him.[3] In addition to the academic philosophical thinking, there are other factors in Borgmann's work. His Catholicism becomes readily apparent to the reader. Moreover, Borgmann's view of the natural sciences is exceptionally benign given his continental background.[4] Borgmann's thinking, furthermore, tends to take place in the presence of wilderness in the tradition of Henry David Thoreau (1817–1862), Aldo Leopold (1887–1948), and perhaps Native Americans. Finally, his thinking and writing are local, clearly North American, and reflect especially the culture of the Rocky Mountain West in the United States.[5]

I begin with a few preliminary considerations that elaborate on Borgmann's notions of science and technology and his understanding of three different discourses in intellectual pursuits. I go on to introduce the device paradigm, the pattern of the technological way to be in the world that Borgmann argues underlies life in technologized societies. Next I look at focal things and practices that constitute the key elements of what Borgmann considers ontological-moral reform of the human life within technology. The end of the chapter consists of a critical assessment of Borgmann's work.

ON SCIENCE AND TECHNOLOGY

It is generally agreed that modern technology has a distinct character at least in part because of a close relationship with modern science. This is a consensus that spans both engineering and humanities philosophies of technology—using Carl Mitcham's terminology.[6] Thus, any philosophy of technology will commonly include some assessment of science and exploration of the science-technology connection.

Oversimplifying somewhat, it is possible to distinguish two basic views of science and the science-technology relationship. Typical of the engineering philosophy of technology is a view of science as providing the truest knowledge of nature, knowledge that is readily available to ground or be exploited by engineering, which itself offers the best means for realizing many if not all human ends. The classic definition of engineering is that it harnesses the powers of nature—as understood by modern natural science—for human benefit. The science-technology relationship, while significant, remains contingent.[7]

Typical of humanities philosophy of technology is a questioning of scientific knowledge as limited in its grasp of the depths of nature, just as modern technology is seen as limited in its utility for human ends. Indeed, modern science is argued to be as much transformed by technological interests as it is an independent form of knowledge, so that its exploitation by technology is more inherent than contingent. The humanities philosophy of technology argument for delimiting technology thus applies equally to delimiting science.[8]

What is remarkable and distinctive about Borgmann's philosophy is that he wants (with engineering philosophy) to interpret science in a realistic sense and (with humanities philosophy) to question technology. The basic argument he provides for grounding his stance is that scientific explanation is continuous with common human experience and reasoning. He uses wine fermentation as an example to show that "the drift to scientific explanation is gradual and inescapable." In other words, Borgmann is asserting that a scientific way of interpreting phenomena is universal for humans. Furthermore, Borgmann contends that reality has a lawful structure.[9] Therefore, it is possible that "a deductive-nomological explanation makes explicit how an event fits into the lawfulness of reality." Through the discovery of new lawful structures in reality and explaining them through subsumption under scientific laws, reality is made more and more intelligible.[10]

Borgmann's scientific realism obviously affects his understanding of the relationship between science and technology. Borgmann thinks that it is possible to distinguish "(1) science as human enterprise, (2) science as the body of well-established laws and theories, (3) science in its applications." An instrumentalist[11] philosopher of science would protest against such distinctions and argue that all three aspects of science—and technology—are integrated and impossible to separate. Borgmann, however, alleges that these distinctions are possible and necessary, since he considers the distinction between science and technology crucial to maintain.[12]

In this light it appears that Borgmann holds a positive view of science. Borgmann, however, observes science in a larger historical and cultural setting as well. Then the role of science in human life does not emerge only as an unproblematic growth of intelligibility. Although he gives science all its due, Borgmann perceives a few troubling cultural questions concerning science. First, he refers to studies that indicate the public's weak command of scientific knowledge. So, although Borgmann grants the best possible status for scientific theories as descriptions of reality, he points out that science as cultural force has problems, since it does not penetrate all the layers of society.[13]

Second, Borgmann notes that some earlier ways to understand the world, such as Aristotelian science, had both world-articulating and world-explaining significance. To articulate, in Borgmann's usage, means "to outline and highlight the crucial features of something." Modern science, however, has abstracted the articulating facet away from its depiction of the world. Therefore, it is not as capable as, for instance, the Aristotelian system of science, metaphysics, and ethics of providing an organic and understandable worldview. The same contrast could be exhibited with the help of most premodern philosophical and religious worldviews, such as Stoicism and Christianity. In the modern period a gap has opened between the exact, explanatory scientific framework and the more insightful, holistic, and value-concerned humanistic-literary approach to reality. This phenomenon is often referred as the split between the two cultures, after C. P. Snow's designation.[14]

Concerning technology, Borgmann outlines three types of theories that can describe technology. The first type he calls the substantive view. It perceives technology "as a force of its own right, one that shapes today's societies and values from the ground up and has no serious rivals."[15] The substantive view, in other words, perceives technology as having a distinct essence and character. Borgmann discusses Jacques Ellul's theory of *La Technique* as an example of this kind of theory. Borgmann's substantive view roughly corresponds to "technological determinism." Substantive, or technologically deterministic, explanations, in fact, do not live only in the books of radical social critics such as Ellul. One can find in high-school history textbooks assertions that the printing press enabled the Reformation or that the steam engine launched the Industrial Revolution. A historical turn is explained, but the origins of the technology that "caused" historical change remain obscure. Thus this approach explains most societal phenomena with monolithic, substantivized Technology but leaves technological phenomena themselves without an explanation. Moreover, its reductionist social descriptions are problematic.[16]

The second type of theory Borgmann calls instrumentalist. The instrumentalist view perceives technology as a value-neutral tool. This approach is congenial with liberal democratic ideas that the state provides neutral means for the good life, but it is up to individuals to decide what the good life is and to strive for the realization of it. The strength of the instrumental view is that it is closer to the common understanding of the role that tools have in society. It also grants a real meaning for human freedom, agency, and responsibility in decisions concerning technology. Yet the separation of means and ends, i.e., the value-neutral concept of technology, is highly problematic. Borgmann summarizes the problems of instrumentalism as follows:

> Any concretely delimited piece of technology can be put forward as a value-neutral tool. But it is a short-sighted view. The availability of mere means is itself a remarkable and consequential fact. Historically, it is just in modern technology that such devices become available. As I will show later, it is equivocation to speak indifferently of tools in a modern and in a pretechnological setting. A means in a traditional culture is never mere means but always and inextricably woven into the context of ends.[17]

The third form of theory is pluralist. A pluralist theory perceives the technological change in all its complex nature, in which material artifacts, technical knowledge, production systems, interest groups, information networks, contrasting points of view, and social relations intertwine. It is a sociological look at technology. According to Borgmann, this view does well in theoretical discussions but does not really describe reality. In reality, he argues, technological change is based on a considerable consensus on "the pattern of technology," not a complex web of contrasting forces.[18]

Concerning his own theoretical attempt, Borgmann says that it should combine the strengths of these three basic perspectives without their flaws. It should have the boldness of the substantive view without leaving the character of technology opaque. It should restore the transparency of instrumentalism without its superficiality. Finally, it should do justice to the nuances of technological activities as social processes without losing sight of the overall pattern of technology.[19]

TYPES OF DISCOURSE

To chart and address the problem of the "two cultures," Borgmann makes a threefold distinction between deictic, apodeictic, and paradeictic discourse.[20] The deictic approach is testimonial in nature; it does not strive for rational cogency or attempt to compel its interlocutor with evidence. Neither does it involve a universal view. Deictic discourse manifests most clearly in poetry, literature, music, arts, and sermons. It takes place at the command of otherness, and it wants to stay open for commanding presence beyond its control. Deictic discourse, in other words, is essentially a non-objectifying approach to things.[21]

Deictic discourse, according to Borgmann, is guided by three elements: enthusiasm, sympathy, and tolerance. Enthusiasm derives from a word that meant "to be filled with divine." In Borgmann's usage, it means that one has been touched by a thing of ultimate concern, which is greater and more enduring than the human being. Simple examples of such things could be the sea for a sailor or the wilderness for a hiker. Sympathy denotes the desire to share the experience of ultimate concern with others. This comes from the observation that this reality and the experiences of it are conducive to human integrity and final well-being. Tolerance comes from understanding that one cannot force one's understanding of these realities on others. One does no more than give witness of one's own experiences.[22]

Apodeictic discourse is the way science approaches reality. It is based on the assumption of the lawful, invariant nature of reality. Science employs experimental and logical methods to achieve the most general level in expressing laws and structures of reality. To be able to do this, apodeictic/scientific discourse approaches reality as abstracted into objects. While deictic discourse dwells in particular on the personal, subjective, concrete, momentary, and unique, science is interested in the general, neutral, objective, abstract, lasting, and typical.[23]

The apodeictic discourse of science is, however, unable to orient human life. Science excels in precision and transparency. But as its method is based on neutrality and its results aim at descriptive factuality and unbiased conceptual structures, science by definition does not say what is important in life, nor does it highlight the essential features of reality. For articulating the latter one needs another kind of discourse. Testimonial approach has the power to speak about orientation, and this is why there is a proper place for deictic discourse in Borgmann's philosophy.[24]

Borgmann's account still needs a third form of discourse. He calls it paradeictic or paradigmatic explanation:

> But there is a third possibility of explaining, one where we try to comprehend *the character* of the reality by discovering its predominant *pattern.* A pattern is more concrete and specific than a law and yet more general and abstract than a unique focal thing. To illuminate reality by disclosing its pattern is a quasi-deictic explanation. Let us call it paradeictic or paradigmatic explanation.[25]

This is the way some social and cultural sciences shed light on their subjects.[26] Forming law-like theories to explain society and human behavior has proven difficult. Some distinctive governing structures, however, appear to exist within the sphere in which the social sciences operate. Borgmann calls these structures patterns, paradigms, or characters. Thus, his own analysis is mainly of this third type: he aims at discovering the pattern of technology and sketching the character of life in advanced industrial societies.[27]

Discovering the pattern of technology is necessary, since science cannot explain technology as a social and cultural phenomenon. Nor does the traditional way of thinking about ethics as such help solve the problem of technology:

> [T]he technological paradigm gives a radically new meaning to goals, values, or ends. If this is overlooked, an examination of technology by way of 'raising the value question' will almost always be ensnared within the framework of technology and fail to be radical.[28]

ANALYSIS OF THE TECHNOLOGICAL CULTURE

Borgmann's paradigmatic analysis of technology concerns distinctly modern technology. It is evident that bronze, the watermill, sails, the alphabet, and the aqueduct were technological artifacts. In fact, the history of technology in this sense is considerably longer than the history of science. However, the cultural transformations that took place in the beginning of the modern age meant the birth of a new kind of relationship between science and technology. This new relationship could be characterized as one of interdependence. Modern technology is something through which scientific reasoning and craftsmanship are combined. In this sense it is a unique phenomenon in the history of humankind. The central element in Borgmann's analysis of technology in this new and historically unique sense is the device paradigm.[29] Using the paradeictic method, Borgmann captures what he claims to be the governing pattern in technologically advanced societies, both at large, at the macro-level of national economies, and at the micro, or personal, level.[30]

Technological Progress under Scrutiny

The historical background of the device paradigm Borgmann calls "the promise of technology." We do not live in a technological world without a good reason. Hunger,[31] cold, heat, storms, earthquakes, darkness, fatal illnesses, material scarcity, and wild animals—in one word, nature—have made human life precarious throughout history, and continue to plague the majority of human beings on the globe. Hence, the promise is simply the following: appropriate knowledge combined with suitable contrivance will free human beings from toil and suffering. This promise was clearly explicated in the early centuries of the modern era by thinkers such as Francis Bacon (1561–1626), René Descartes (1596–1650), and Tommaso Campanella (1568–1639).[32] The promise retains its allure both explicitly and implicitly, and is usually articulated when new technological innovations are introduced in society.[33]

Looking back in history, it is clear that the promise of technology in the above-mentioned sense has been fulfilled. It was not religious prayers, moral striving, love for beauty, or even pure science itself that rescued humankind from harsh conditions. It was the application of technological innovations like the sewer, the steam engine, electricity, the assembly line, and central heating.[34]

It could obviously be argued that technological devices that have helped our culture achieve well-being were essentially applications of science. To such a claim it is possible to make three responses. First, the history of technology shows that early-modern technological innovations, such as the steam and combustion engines, were made independently from scientific theories. Indeed, scientific theory often comes afterward and explains a technical development—as thermodynamics developed to explain the increased efficiency of Watt's steam engine. Second, there is no good reason to claim that technology is dependent on science but science is not dependent on technology. Consider, for example, the telescopes and particle accelerators needed in modern physics. A third argument could be based on the fact that the history of technical devices is far longer than that of science.[35]

To return to the fulfilled promise, it has larger social and political implications beyond the immediate reality of natural dangers and toils. Good grounds exist for thinking that without modern technology, liberal democracy and the ideas of liberty, equality, and self-realization could not be actualized en masse.[36]

Borgmann claims that this is exactly what the promise of technology is all about: according to him, technology actually promises to grant us the good life, a better life, better in the sense of being enriched, happy, excellent, conscious, and wise. But here he identifies a problem. Technology has provided material security and well-being for many, but has it made our lives better? Borgmann makes use of social scientific studies and statistics to show that it appears, within the precision with which these kinds of questions can be answered, that our lives have not become unqualifiedly better in the course of technological advance.[37]

To explicate the good life in this context Borgmann formulates an ideal of excellence reflecting the classical and Judeo-Christian roots of the Western tradition:

> (1) The excellent person is a world citizen who understands the structure and coherence of the universe in its scientific and historical dimensions. (2) The excellent person is gallant as well as good and intelligent and seeks physical valor as well as intellectual refinement. (3) The excellent person is accomplished in music and versed in the arts. To the virtues of world citizenship, gallantry, and musicianship the Judeo-Christian tradition has added (4) the virtue of charity according to which real strength lies not in material force or cunning but in the power to give and forgive, to help and heal. These kinds of excellence were periodically revived and reshaped by such movements as chivalry, humanism and the Renaissance.[38]

In light of the ideal of excellence Borgmann points out numerous shortcomings of our culture. Although the body of scientific knowledge expands, this increase in knowledge translates poorly into more enlightened citizenship. Despite nutritional knowledge and possibilities for exercise, obesity is an ever-growing problem, and the physical condition of the population is degrading. Divorce rates tend to rise the more developed the society becomes. The public awareness of communal, national, and global issues does not appear to improve due to economic growth, technological development in communication devices, and increased leisure. Voter turnout in elections is generally decreasing. Nor do people spend their free time learning arts or new skills. Borgmann's idea is not to compare our culture with some earlier epoch, but to evaluate our culture's achievements against what appear to be its inherent ideals and possibilities.[39]

In other words, if one examines in the mode of paradeictic characterization the tradition that has produced both the promise of technology and the ideal of excellence, it appears evident that there is a conflict between the ideas of excellence and the present reality. It appears that those earlier thinkers who articulated promising aspects of technological advance had the realization of some form of excellence in mind, not stultifying leisure.[40]

Obviously, happiness and excellence are themselves ambiguous concepts and difficult to study empirically. Some studies are, nevertheless, informative, and a few conclusions may be drawn from them.[41] For instance, from the studies made on TV watching, one can discover that while the proportional time spent in front of the television set has been steadily increasing,

TV watchers have not found this pastime in any way excellent. The term "couch potato" reveals this judgment.[42] Thus it appears that people's actual ways of life and their deepest convictions are in discord.

The Device Paradigm

Now we have arrived at the question to which the device paradigm is an answer. Borgmann argues that life has been and is being given a distinctive structure due to technological revolutions and devices that have permeated our culture and way of life.[43] One telling example concerns warmth. In premodern days, warmth was produced by a hearth. It kept the temperature in the house comfortable and provided heat for baking and cooking. An old-fashioned stove, however, required care as well. Someone needed to select, saw, and split the wood, carry it in, and keep an eye on the fire in order to maintain a suitable level of heat. All this required skill and created a small social network around the stove: the tasks needed to be divided within the family. The hearth was also a particular thing, a metal or stone construction usually placed in a central corner of the house. It was a place where the family gathered for warmth, for food, and often just to chat. This may sound romantic, but taking care of the hearth was burdensome, too, and sometimes a real hassle. There could be a shortage of wood, the wood might be green or wet, quarrels would occur about the sharing of jobs, and in any case numerous chores needed to be taken care of.[44]

Today our houses no longer have hearths—except as decorative imitations. We have been disburdened from the tasks of heating. Most often this has happened through a gradual development that resulted in the installation of a central heating system. Normally, we do not even know where the heating plant is located; we merely receive the commodity it produces: warmth. There is no hassle, no smell of smoke, no metal object, and no social relations, either. This is an actualization of the device paradigm.[45]

The device paradigm refers to the general pattern of the technological way of taking up with the world. The following aspects characterize the device paradigm: disburdenment, disengagement, decontextualization, availability and disposability of commodities within it, and discontinuity.[46] Technology disburdens us by providing commodities that are available without effort or skill, any time, anywhere. If one wanted to enjoy Mozart's *Don Giovanni* in Copenhagen in the 1840s,[47] one had to find a theater where the piece was played, wear suitable clothing, buy a ticket, be present in the right place at the correct time, and face the social situation of being at the opera. Under contemporary conditions, we have witnessed innumerable phases of development in analog and digital audio technology and the advances in video production and distribution from motion pictures to high-definition television. Consequently, a contemporary Dane can enjoy *Don Giovanni* with CD, mp3, or DVD equipment and a surround-sound system, beginning with a favorite aria.[48] Technology disengages and decontextualizes the desired commodity—warmth or music—from all the other aspects.

By discontinuity, Borgmann means the tendency of the device paradigm to create sharper polarities than the premodern ways to encounter reality. The means-ends distinction is the most characteristic.[49] In taking care of a wood stove, several ends were involved. Moreover, means and ends were integrated or interwoven. The central heating system produces only one commodity. This commodity is a mere end; we do not even know clearly how the heat is produced; the machinery is hidden, it is mere means. In going to the opera, several ends were interwoven with several means. Listening to a CD produces a single end, the music, effortlessly, while concealing the mere means, the electronic machinery for reading the digital information.

The sharp means-ends distinction is connected to what Borgmann calls the separation of a "foreground" and "background" in technology. The background of technology presents itself in our lives as work and labor. The processes of production and organization in contemporary society are immensely complex.

> It leads to devices that are physically so dense and complex that they are beyond direct intervention or modification by anyone. Thus crucial and growing parts of the technological machinery are becoming intellectually and physically ever more remote from the common person's competence.[50]

Not many people, for instance, would know what are the processes needed for a cell phone call. But not only machine-like gadgets follow the pattern of technology: administrative and economic systems function in a similar manner.[51] A professional person today is an expert in a tiny sector of some field: pension insurance systems, production processes of a certain part of a cell phone, foreign policies in Southeast Asia, and so on. A blue-collar worker usually executes a small part in a complicated production process. Thus we go to work in the background of technology, and there we meet with one small part of the socioeconomic system. At least in the beginning of the industrial era, this kind of work used be monotonous and unrewarding for the majority of citizens. More recently, the character of labor has taken new forms due the transition toward the service or information society. However, the work in an office cubicle or in a sales position does not appear less unrewarding or monotonous, although it has been disburdened from physical labor.[52] This is the means side of life: for the worker the means to obtain money, for the employer the means to obtain human resources.[53]

The end side is leisure. When we spend our free time we encounter the foreground of technology. This mostly consists of commodities that are available for quick consumption, easy use, that conceal their means side beneath a glamorous cover.[54] The cell phone is a good example: the electronic hardware is physically hidden in a cool stylized cover and the complex software is concealed beneath an easy-to-use operating system, within which one can handle the calendar, games, photos, music, videos, and more. The foreground of technology requires few skills, no commitment, no effort from us; it is a matter of buying and using.[55] A person can purchase clothes that build her identity, TV programs, vacations, bank and insurance commodities, different medical and therapy goods, ready-made food, sports entertainment, education, daycare for children, legal drugs, beauty enhancement, a bungee jump, the experience of being a world leader in a strategic computer game, and so on.

> But it is clear that, if technological boredom is to be kept at bay, there must be a kind of enrichment which is unlimited. It is available under the heading of entertainment and comes in the shape of commodities that we ingest, that we eat, see, or hear. These constitute the final and central commodities and the foremost foreground of technology.[56]

In such a manner the device paradigm has sharply divided life into work and leisure. In the same way, the distinction of body and mind is sharpened since the disengaging tendency in technology, through physical disengagement, is most often also a disembodying tendency.[57] Going to the opera had its bodily dimensions, but listening to a CD can be an act of ears and mind along. Moreover, the gym workout is targeted to a lonely body, whereas boxing or manual labor involves mental and social dimensions. Obviously, one often listens to music through earphones in the gym, but that even more separates the mind from its bodily context through its focus on a musical world originally produced in some unknown studio. Thus there are good grounds for claiming that technology strengthens the opposition of individual and community, since nearly all technological commodities allow solitary consumption—and sometimes promote it—while pretechnological cultural forms were more collective in their nature.

Furthermore, Borgmann notes that there is a growing tendency to address the members of a technologically advanced society not as citizens but as consumers. Consumers can affect the world only through their buying choices, whereas the concept of a citizen—member of a political community—implies the right, knowledge, and power to discuss fundamental principles.[58]

In *Technology and the Character of Contemporary Life* Borgmann concentrates mainly on the contrast between things and devices in a rather tangible sense, and follows the existential, cultural, and political implications of the device paradigm. In later works, especially in *Crossing the Postmodern Divide* and *Holding On to Reality*, he has widened this analysis toward considerations of reality. This expansion follows the development of the device paradigm from a slightly different angle, and pays special attention to recent innovations in information technology. However, Borgmann often emphasizes that these developments are nothing entirely new, but on the same continuum as the device paradigm:

> The progress of technology, once its basic pattern is seen, has had a numbing sameness for at least a century. There is nothing in the most recent innovations and prospects that falls outside of the pattern of technology.[59]

The main point in Borgmann's examination of hyperreality[60] is that technology creates a substitute for the physical or natural reality.[61] He analyzes at length the ways in which information technology and different forms of simulation present parallel worlds to the primary one. However, hyperreality in Borgmann's sense can be also found in simpler surroundings. For example, technologically modified foods present instances of hyperreality. In the natural reality, creamy and sweet desserts are enjoyable, but because of their constituents they pose problems for health and weight control. With modern food-processing technologies, it is possible to produce sweet soda drinks without calories or products looking and tasting like whipped cream but devoid of cholesterol and with few calories. According to Borgmann, these kinds of groceries are cases of tangible hyperreality. They present a feature that Borgmann attributes to hyperreality: it is flexible for the user's desires. If I want to enjoy cake but avoid getting fat, such groceries fulfill my desires. This shows that they illustrate the device paradigm as well, since the idea behind them is the separation of taste qualities from nutritional elements. The complex biochemical compound of the grocery product is the background in this case, and the entertaining sweet or creamy taste the foreground.[62]

Socio-Ethical Considerations

The technological way of taking up with the world through the expansion of devices has troubling aspects and consequences, but this obviously does not mean that technology in itself is bad or that it has not provided us with numerous helpful instruments. The problem is more how individuals, communities, and societies could have a different relationship to technology—as material structures and as the way of handling the world. Now they appear to lack the awareness of the life-forming role technology has in our culture. Thinking about technology has been ambivalent, a vague mixture of contrasting understandings of technology. It often happens that we perceive technological devices and systems simultaneously as neutral instruments and as providers of the good life. According to Borgmann, this lack of awareness is a problem of current ethics and political philosophy as well.

Individuals, communities, and societies face a significant problem of orientation due to technoscientific advances. As was noted in the discussion concerning different discourses, scientific discourse does not provide assistance in questions of orientation. Yet science has made obsolete many traditional sources of orientation, which are now considered mere myths. Technology, by contrast, has a certain orienting force, but it stays almost completely out of reflective sight or critical conversation. This is exactly the problem that Borgmann attempts to analyze through the characterization of the pattern of technology. Although science does not provide them, directions for orientation are needed. Borgmann observes that a highly appreciated answer for the needs of orientation is "value talk," "raising the value question."[63]

Borgmann, however, finds this solution problematic. He argues that, in contemporary culture, "values," like nearly everything else, are seen through the device paradigm.[64] "No matter how the question of values is raised and settled, the pattern of technology itself is never in question."[65] So, unless the discussion can reach beyond the device paradigm, the values that are weighed and debated are treated like commodities. In other words, they will be seen as ends, more precisely as mere ends, within the sharp means-ends distinction of the device paradigm. Further, using Borgmann's idea of the background and foreground of technology, it would be possible to analyze how the concept of "value" is split between them. On the one hand, it appears that in some contexts values are chosen very freely, based on arbitrarily subjective choice. An individual may choose and profess any kind of values: safety, family, work, health, green values, adventurousness, religious values, and pleasure, for instance, can be personal value preferences. Even if someone professed a personal value of experiencing pain and decided to sit with some thumbtacks on her chair, it would not be easy to point out that this value is somehow strange. On the other hand, in other contexts there appears to be very little or no room for choice due to economic "necessities." For instance, the marginal, within which "the environmental values" can be pursued, is usually narrow and clearly defined in money. The light values are in the foreground of technology and the grave, tightly conditioned values are in the background, on the side of mere means.[66]

Thus it is also obvious that values are understood as abstractions that are disembedded from larger contexts and concrete things. Concrete things are rich in features and subtle meanings, but values are more like quantitative functions, abstract concepts of the goods of life. In pretechnological times moral meanings were closely tied to social classes, livelihoods, and comprehensive ways of life. Modernization has made values more abstract, as both technology and social structures have enabled many value combinations that would have been self-contradictory earlier. Borgmann admits that value discussion has a role in society, but demands that its shortcomings be noticed and that it not conceal the more fundamental question of the device paradigm.[67]

In his 1984 book, Borgmann directs criticism on his contemporary—the 1970s and '80s—Marxist political theories in a similar vein. He perceives that while emphasizing economy and power relations, Marxist theoreticians fail to attribute to technology the role it has in the society. Further, he notes that the particular problem of Marxist theories in postindustrialist society is the difficulty of identifying the ruling and ruled classes. This kind of class antagonism is an essential idea in Marxism. Therefore, the absence of it will seriously undermine the fundamentals of Marxist theories.[68]

But Marxist theories were not in the mainstream even in the 1980s. Mainstream political philosophy in advanced Western societies has for some time been in the liberal democratic tradition. Borgmann notes that liberal notions of liberty, equality, and self-realization appear to be happily in harmony with technological advance. This is especially true if technology is understood in instrumental terms. Oversimplifying somewhat, in liberal democracy the government takes care of guaranteeing maximal liberty among citizens in a manner that should protect their equality. The government, however, is supposed to favor any concept of the good life as little as possible. The pursuit of the good life, and happiness, belongs to the domain of self-realization, and it is perceived as a matter of individual choice. It is said that the question of the good life is left open in the liberal democracy.

In the context of these liberal democratic notions, discussions of the good life become difficult and touchy. On the one hand, there is no need to discuss the good life, since it is everybody's private business. On the other hand, human beings are social creatures, and a totally private good would be an impossible idea.[69] So, there is a need to discuss the good life, but at the same time the danger of paternalism curbs the attempts at that discussion. But nothing is lost, since the question is still left open.

Borgmann, however, argues that this is an illusion. The question of the good life has already been silently answered by technology. This answering takes place in the structuring of "the material culture." The term refers to the human significance of the human-built world.[70] As elaborated earlier in this chapter, the device paradigm has organized life in a distinct manner. From this perspective it is no wonder that the discussion of the good life is absent: it is an "illegitimate," politically incorrect topic, and there exists a persistent misunderstanding concerning the openness of the question. Therefore, the way to a discourse on the good life goes through correcting the misunderstanding:

> It appears then that we can make room for the discussion of the good life only if we have before us the pattern of technology and the peculiar way in which, *under semblance of a mere tool*, it preforms our discourse, and agenda.[71]

Borgmann's arguments as reviewed here force one to return to the claim that we cannot say that technology itself is bad or good. Arguments on these lines appear to fall prey to the above-mentioned illusion. Obviously, technology in itself cannot contain moral qualities. But it is not fully neutral either.[72] One must ask: If we as communities consciously pursued the good life—according to some understanding of it—instead of thinking that all technological solutions, from gadgets to organizational structures, are morally neutral, would not we design different kinds of technologies? The answer to this question is hypothetical at present. Borgmann, nevertheless, makes it clear that it is not possible to think about technological development along neutral lines:

> Some observers of modern culture do note disparities and tensions between contemporary doing and making, between ethics and technology. They complain that, while technology has advanced dramatically, our moral aptitude in dealing with it has not. But this complaint is as radically mistaken as the general divorce of doing from making. It fails to see that a technological accomplishment, *the development and adoption of a technological device always and already constitutes a moral decision.*[73]

APPROACHES FOR A REFORM

If all Borgmann's concerns were summarized in one word, the word could be the "unfocused" or "disintegrated" character of contemporary life. Consequently, the cure should lie in restoring a focus to life. Indeed, Borgmann's key concepts in the reform of life in technological societies are "focal things and practices."[74] Focal things and practices are at the center of this section, as it looks at Borgmann's reform proposal for technologized culture. The first part of the section treats in detail Borgmann's ideas of reform from the viewpoint of everyday material culture, while the latter follows how Borgmann expands his ideas toward ethics and wider societal settings.

Things and Practices

The word "focal," used repeatedly by Borgmann, derives from Latin *focus*. *Focus* originally referred to the hearth in a house. Thus Borgmann's example concerning warmth has one more motif behind it. In a Roman house, *focus* was the center of the home. Meals, sanctification of marriage, social life, and invoking the house gods took place before the hearth. In early periods the dead were even buried by the hearth.[75] In modern times the word "focus" has received more abstract meaning:

> [A] focus gathers the relations of its context and radiates into its surroundings and informs them. To focus on something or to bring it into focus is to make it central, clear, and articulate.[76]

Borgmann thus refers to things that would have a centering power that could be, at least in a metaphorical sense, something like what the hearth was to the Romans.

Borgmann's idea, in short, is that we look in the wrong place if we expect technological disburdenment to provide the good life. Instead, what technology can do is free us to enjoy things and practices in their primordial sense. To be able to do this we need to take seriously the disappointment connected to the irony of technology, and to see through the pattern of technology:

> Heidegger says, broadly paraphrased, that the orienting force of simple things will come to the fore only as the rule of technology is raised from its anonymity, is disclosed as the orthodoxy that heretofore has been taken for granted and allowed to remain invisible. As long as we overlook the tightly patterned character of technology and believe that we live in a world of endlessly open and rich opportunities, as long as we ignore the definite ways in which we, acting technologically, have worked out the promise of technology and remain vaguely enthralled by that promise, so long simple things and practices will seem burdensome, confining and drab. But if we recognize the central vacuity of advanced technology, that emptiness can become the opening for focal things.[77]

Borgmann and Heidegger[78] are interested in things, since the device paradigm tends to do away with things in their deep sense, in their "thing-ness." The example of the hearth indicated how a thing, an iron or stone stove, was replaced by a system that is not really present as a thing. In the same way, fast-food production systems and TV dinners replace the kitchen utensils and dishes involved in a family supper with paper cups and plates. Within the device paradigm things are not really things but disposable parts of a system.[79]

> Commodities, in comparison with focal things, are highly reduced entities and abstract in the sense that within the overall framework of technology they are free of local and historical ties. Thus they are sharply defined and easily measured.[80]

Consider further concrete examples. The wheelwright's craft one hundred years ago had to do with focal things, cart wheels, whereas work in an assembly line constructing car wheels is more an actualization of the device paradigm.[81] Fly-fishing can be a focal practice, and then the fishing rod and/or the trout is the focal thing.[82] For a runner the running route is a focal thing, for a backpacker the wilderness. I suppose, however, that activity does not necessarily require running speed to be focal. If that is true, then simple walking, as described by Henry David Thoreau in his essay "Walking," can be a focal practice.[83] Moreover, arts and crafts can be focal practices.

Further, a family dinner is a focal practice and all the material objects involved in it are focal things within that practice; fast food and ready-made meals are its equivalents within the device paradigm. Borgmann in fact discusses a "culture of the table" that can involve a lot more than just cooking and eating: it has a beginning and an end; it is a tangible way in which the family comes together. It is in dialogue with traditions of cuisine, and in religious families it involves grace through which one can express gratitude for the gifts of the earth.[84]

Consider now focal things and practices from an analytical perspective. Borgmann remarks that while dealing with focal things one is guided by a commanding presence that is exactly opposite to the distraction produced by devices. Focal practices require exertion, skill, self-transcendence, perseverance, endurance, patience, commitment, and attention—qualities from which the leisure side of the device disburdens us. In this way, Borgmann thinks focal practices can recenter our lives and provide contexts in which virtue and excellence can be worked for. Discontinuity characterizes the activities within the device paradigm, whereas focal practices lead to unification and integration: in fly-fishing body and mind work together; in running the runner merges with her environment; in backpacking culture, nature, hiking traditions, and the present converge. The same applies for means and ends: in most focal practices they cannot be strictly separated—running is simultaneously an end in itself and a means to maintain well-being: "In the runner, effort and joy are one; the split between means and ends, labor and leisure is healed."[85]

Focal things are closely related to the respective practices. In this sense they are things in the human world, things culturally appropriated. Moreover, as already noted, the technological environment provides a contrast that augments the meaning and depth of focal things.[86] Within the device paradigm everything is controlled, everything is made predictable; in contrast, otherness lives within focal practices, wilderness being one of the best examples.[87]

> [W]e can hope to speak in a principled and forceful way about the good life when we allow ourselves to be guided by focal things, matters of ultimate concern that are other and greater than ourselves. Such a discourse, which I have called deictic throughout, does not strive after cogency since it cannot, nor does it wish to control its subject matter. But neither is it arbitrary since it is guided by an eminent, publicly accessible, and tangible concern which can be pointed out and explained.[88]

So, for Borgmann, speech concerning the focal practices is testimonial in character; it takes place in deictic discourse. One cannot force others to recognize focal things. Yet if one has witnessed the reorienting power of a focal thing one is able to share this discovery.[89]

In the preceding section I touched upon Borgmann's interpretation of hyperreality. In conjunction with this critical analysis, Borgmann maintains that one can encounter more "real" reality amongst the focal things and practices. In some sense the way to focal reality is the middle path between the passive consumption and hectic "hyperactive" working life. Whereas hyperactive overachieving strives for maximum control, is overly futuristic, and dwells within abstract concepts and artificialities, focal realism is open for things and events, stays calmly in the present moment, and is interested in concreteness in all its nuances. Focal realism, however, is not dreamy romanticism, but requires "patient vigor." Therefore, it appears active in contrast to the passive consumption of hyperreal products in the foreground of the device paradigm.[90]

Focal realism, in addition to having contact with primary reality, is "realistic" in a practical sense. Borgmann notes that because we live in a finite world, there is only limited time and place available. This means that if TV watching takes more than three hours of a person's leisure, other activities will be deprived of those hours. This is a truism that is surprisingly often forgotten in the technological mass culture, where all new devices are seen as additions to everything else that is already possessed. For example, when information technology is proposed as a key for better education it is seldom assessed what is pushed aside by the new devices and structure of classrooms. The simple fact, however, is that all the time, space, and attention given to new devices have to be taken from other things and activities. And the other way around: if one wants to cultivate focal practices, it means reducing the time spent with devices. Therefore, patience, resoluteness, and vigor are needed if one wants to break away from the rule of the device paradigm.[91]

As Aristotle notes, "one swallow does not make a summer";[92] one focal act does not make a practice. Nor does a singular omission of focal action dismantle a focal practice. An act and a practice are different, and establishing a practice has far more moral weight[93] than just committing one act in the domain of respective practice.[94]

In fact, I read Borgmann as wanting to scrutinize larger moral settings instead of instantaneous decisions. He is saying that the good life is essentially not a collection of separate good deliberations on deeds. The good life can mainly be strived for through reflecting on the ethos of a person's or a community's life, critically assessing the material culture by which one is surrounded, and the molding of regular practices. In other words, he argues that more attention has to be paid to the choices concerning these larger structural settings and the temporal character of morality.[95]

Yet the commitment to practices is difficult, not only because of outside distraction, but due to their inherent dynamics as well. Borgmann's sources on the different practices report on "the dark night of the soul," emptiness and depression, that can settle exactly at times when one has been most dedicated in the practice.[96]

One way to elaborate Borgmann's ideas on focal reality is to take some distance and look at them next to existential philosophy, in particular Heidegger's early philosophy in *Being and Time*. In Heidegger's description, *Dasein*'s being in its everydayness is always more or less fallen into inauthentic (*uneigentlich*) existence, in which it avoids facing the finitude of life and its own personal possibilities.[97] Heidegger's account of how *Dasein* may strive toward authenticity (*Eigentlichkeit*) through accepting the primal anxiety, listening to the call of conscience, and running-ahead-into-death includes passages of intense philosophical analyses.[98] However, in his depiction of the possibility of authenticity, Heidegger concentrates on the inward life in a highly abstract manner, although his picture of inauthenticity involves some references to more concrete phenomena such as press and radio. One way of understanding Borgmann's focal philosophy is to see it as a concrete description of the conditions that can lead toward more authentic existence. In this sense it may be read as a reflection on Heidegger's early philosophy with the help of his later concepts.

Reform on a Societal Scale

I will end this section by looking at Borgmann's reform proposals from a societal perspective. "To make the technological universe hospitable to focal things turns out to be the heart of the reform of technology," as Borgmann sums up his goal.[99] In the following I will raise some crucial issues related to this project.

First, the role of focal things in Borgmann's theory is not only to provide a better sphere of living than that provided by the device paradigm, but also to enable people to weigh and assess technology. It easily happens that we, as individuals and as a culture, forget other ways to be in the world when the device paradigm becomes all-pervading. Interaction with focal things can, however, help one to open a new perspective into life in the technological world.

> A focal practice engenders an intelligent and selective attitude toward technology. It leads to simplification and perfection of technology in the background of one's focal concern and to a discerning use of technological products at the center of one's practice.[100]

In this sense Borgmann calls focal things and practices "metatechnological things and practices."[101]

Second, the societal viewpoint on things and practices must address the question of the plurality of focal concerns. In medieval society there was one—institutionally reinforced—unifying ultimate concern, which affected the whole culture from politics to art. In modern societies there is no such shared highest good.[102] In this context different concerns inevitably end up competing for the time and attention of the public, and within finite individual life there is room for only a limited number of focal practices. Therefore, there is a need to assess focal practices and for demarcation of them as well.[103] As a possible method for the assessment of focal things and practices, Borgmann discusses John Rawls's Aristotelian Principle:

> [O]ther things being equal, human beings enjoy the exercise of their realized capacities (their inner and trained abilities), and this enjoyment increases the more the capacity is realized, or the greater its complexity.[104]

This principle, however, is partly problematic in the context of Borgmann's focal things. Realizing capacity fits his argument well. But computer programming—and in some cases even gaming—at first sight is far more complex than running, cooking, or fly-fishing. Thus, complexity in its mainstream meaning does not do justice to the sensitivities that Borgmann perceives as essential in focal practices. Therefore, he introduces the concept of engagement as a balancing supplement to complexity:

> [T]he acquisition of skills, the fidelity to a daily discipline, the broadening of sensibility, the profound interaction of human beings, and preservation and development of tradition. These traits we may bring together under the heading of engagement.[105]

Hence, the method for assessing the focal practices is an equilibrium between Aristotelian Principle and the viewpoint of engagement.

The discussion on ways to compare and assess practices is surely important. Yet it appears obvious that people with different focal concerns can easily appreciate each other's practices. Runners might not want to raise their practice into a universal position, although it can be the deepest concern for them personally. Moreover, experience in engaging in one focal realm will probably make a person able to perceive the depth of other realms. For instance, artists from different art forms often can see the depth and meaning in each other's work.[106]

Third, with his contrast of concepts Borgmann gives a deeper meaning to the idea of quality in life. Standard of living is characterized by quantitatively advancing economic machinery and glamour in the appearance of the technologically organized lifestyle and its devices. Borgmann calls this affluence. He wants to set an alternative socioeconomic goal to challenge the one measured by standard of living. This he defines as wealth or quality of life.[107] Quality of life is something in which there is room and real possibilities for focal things and practices within the device paradigm.[108]

> Public policy tends toward a reform of technology when it guards and nourishes things in their own right, those which we do not consume but meet with respect and engagement. Setting aside wilderness areas, keeping rivers free-flowing, rehabilitating lakes, saving endangered species, promoting the arts, literature, and the useless sciences, supporting education as a means of achieving world citizenship and the excellence of an engaging life—these measures advance the quality of life appropriate to wealth.[109]

Borgmann has ideas about what the reform might mean to society at large. He draws his political suggestions from liberal democratic tradition: he wants to keep legality and morality clearly separated. At the same time he thinks that goals such as security, equity, peace, and ecology are not "thick" enough content for the good life. They are more like necessities, but without sufficient infrastructure. Therefore, the decisions concerning the good life have to be informed and inspired by public discussion, which takes place in the deictic register. Borgmann, however, notes that there is a tension in American conservative thinking, since on the one hand it stresses familial "moral values," but on the other hand it makes economic decisions that threaten the integrity of the family.[110]

His most important political-economic structural idea Borgmann borrows from John Kenneth Galbraith. Based on Galbraith's studies, he argues that the economy already has two distinct sectors and will evolve even more in this direction—especially if so steered. The first sector is composed of a relatively small number of gigantic corporations; the second is formed by a huge number of small, local private businesses.[111] Borgmann's large-scale reformation suggests that politics should reinforce the small-firm side of the economy. This would aim at creating two distinctly different spheres of the economy. The first would be local and labor intensive, and it would provide, for instance, food, furniture, clothing, health care, education, and instruction in music, arts, and sports. The second would be based on mass production and automatization, that is, on the device paradigm. It would produce infrastructural goods in areas such as transportation, communication, machines, raw materials, insurance, research, and development. In this kind of two-sector economy, the device paradigm and the focal things would have been set in their proper places.[112] The following quotation summarizes Borgmann's understanding of the reform:

> [Reform] is rather the recognition and restraint of the paradigm. To restrain the paradigm is to restrict it to its proper sphere. Its proper sphere is the background or periphery of focal things and practices. Technology so reformed is no longer the characteristic and dominant way in which we take up with reality; rather it is a way of proceeding that we follow at certain times and up to a point, one that is left behind when we reach the threshold of our focal and final concerns.[113]

ASSESSING BORGMANN'S PROJECT

I will now consider Borgmann's approach critically. I will begin by looking at the role of worldview in Borgmann's approach, which will be followed by a discussion on Borgmann's realism. I will then proceed to discuss demarcation of focal things and practices. In the third section I will look at the implicit understanding of human being in Borgmann's philosophy of technology. Then I will continue to review the criticisms of Borgmann's concept of technology. The final section will make a cultural comparison of focal things and practices in traditional Japanese culture.

The Good Life, Ideology, and Things

In relation to the ideas concerning the good life, Borgmann is simultaneously loyal to liberal democratic thinking and critical of it. His loyalty becomes visible in his reluctance to advance any clear "ideological"—in the sense of being derived from some thought system or religious doctrine—view of what the good life is. His criticality finds its articulation as Borgmann argues that leaving the good life out of the public discourse is untenable. "Minimally" read, he can be understood to be saying: whatever your idea of the good life is, the role of the devices and things in your life is going to affect your pursuit of the good life. Hence, in this "thin" reading his thinking concerns the conditions of the good life, not its definite, "thick" content. Slightly more strongly, he can be understood to mean: life in which focal things are in place is better than the life where they are abolished. Then, at his strongest, he has contentual, ideological views concerning the good life: it is active, artful, and open to nature, reveres traditions, and is in some sense religious, for instance. I think Borgmann is at his best when he analyzes how things create the processes that make life better. I would call this his non-ideological view of the good life. It is, however, impossible to pursue the good life without some thought-through account of worthwhile goals in life, that is, without an ideology. So, the good life does not survive just with things without an ideology. Borgmann is most intriguing when the tensions between the thing-driven and the ideology-driven good life become visible in his work. Then, too, he is most vulnerable to criticisms that find him romantic, conservative, or an advocate of Catholic ideas.[114]

Considering Borgmann's Realism

Tensions appear to exist in Borgmann's notions of realism. To begin, I am not completely convinced that Borgmann's scientific realism is compatible with his Heideggerian view of technology and its role in the human condition. Borgmann, after acknowledging a distinction between instrumental and realistic interpretations of modern science, comes down firmly on the side of realism: "Science in its central sense gives a representation of the general structure of reality that, for the first time in human history, exacts universal assent."[115] At the same time Borgmann argues that technology as devices distorts the human experience of reality. Certainly a conceptual distinction can be made between science and technology. I am not sure, however, whether this unbiased science and dubiously biased technology can be kept separate in concrete reality.

A second doubt about Borgmann's understanding of science arises from his claim that scientific thinking is continuous with a primordial human approach to the world. "Scientific explanation is not a novel assault on the world but the radical precisioning of a procedure that is as old as humanity."[116] This is obviously a line of argument that a scientific realist must follow. In Borgmann's own tradition, Husserl and Heidegger, however, saw the situation in a rather different manner. They thought that the pre-scientific relationship to the world is fundamental for human life and that scientific abstractions distort that relationship. In fact, the phenomenologists intended to go "backward" from "the natural attitude" to an even less abstract way of experiencing the consciousness of the world rather than building more theories through science.[117] Against the background of a phenomenological analysis of science, Borgmann's continuity claim sounds awkward. He appears to seek a view that would balance both the primary role of the lifeword (*Lebenswelt*) and the realistic understanding of scientific theories.

Furthermore, there appear to be good reasons to blame not only the technological device paradigm but also science for the loss of focal things. The scientific attitude is one aiming at detachment. It is therefore at odds with engagement and the commanding presence of things. It is difficult, for instance, to imagine how a botanist—while practicing her discipline—might experience the otherness of wilderness. Since her discipline requires projecting a classifying framework on nature, it is not evident that the practicing botanist would be simultaneously open to the otherness and let the things just shine forth. Further, the analytic tendencies of the sciences appear to contribute to the development that reduces things to mere objects, abstract measurable entities. This again is Heidegger's view in his essay "The Thing."[118]

Maybe a more fruitful way to put this criticism is to go through Heidegger's understanding of truth. If truth is understood as *alētheia*,[119] disclosure, then it follows that every way to disclose being is simultaneously a closure that excludes some other ways to disclose. When science and technology become dominating ways to disclose being, other ways threaten to lose their status as approaches that lead to the real.[120] Moreover, in this development, culture and its members begin to drop the skills—reflective thinking and various crafts, for instance—required for other kinds of disclosures. Oversimplifying to an extent, this appears to be Heidegger's worry.[121]

These are interesting questions, but maybe it is somewhat demanding to require thorough-going treatment of them from a thinker whose primary interest is the impact of technology on human existence. Moreover, the criticisms concerning Borgmann's scientific realism are not fatal, since Borgmann's analysis of technology and the good life is not affected by whether or not he is a scientific realist. One cannot help but suspect that Borgmann's scientific realism is to some extent a rhetorical stance taken so as not to distract readers, who are themselves by and large realists, from his larger argument. Furthermore, as it sometimes appears that continental philosophers do not really give science its due, Borgmann's approach to science is in this context independent and respectable.

Humans in Technology

Consider next Borgmann's anthropology. It appears that he has a rather optimistic view of human beings. He sincerely believes that couch potatoes will rise and leave their remote controls once they notice the rich and deep life of hiking. But this does not happen easily. Already Aristotle had a hard time understanding *acrasia*,[122] the human weakness of the will or incontinence, and the Catholic church defined gluttony, lust, greed, and sloth as deadly sins well before the device paradigm. I think Borgmann, in part, is facing a timeless problem—not one that is unique to technological society—when he is trying to persuade his contemporaries to voluntarily give up disburdenment.

Borgmann's solution significantly resembles Aristotle's, whose idea is that people make a mistake if they seek pleasure in the easy consumption of passive bodily pleasures like eating. He says that the real pleasure is in activities (praxis) that one is good at. These practices are ends in themselves.[123] One could propose that Borgmann's theory is exactly the Aristotelian theory of pleasure and *acrasia* modified for the modern technological context. Borgmann wants to show his contemporaries that they make a mistake when they take TV watching and fast food as pleasurable, since the real enjoyment lies in things with commanding presence. In this context one could simplify and say that according to Borgmann devices work for *acrasia.*[124]

As a last comment on Borgmann's view of human beings, consider the role of things in his philosophy. Borgmann is concerned with the loss of things, but he seldom discusses what happens to human relationships in the technological world. Obviously, such manifestations of the device paradigm as TV or personal music players, and such focal practices as the culture of the table, have to do with restructuring the patterns of human relationships. However, in this respect Borgmann's philosophy echoes Heidegger, who wrote comfortably about a hammer, a clay jug, and a silver chalice, but whose descriptions of social life are somewhat grey and emphasize other people as a threat to *Dasein*'s authenticity. By contrast, for instance, Martin Buber's (1878–1965)[125] existential philosophy grants a central place for personal relations. According to him, human beings can have two principal relationships to the external world: open, personal I-You (*Ich-Du*) and demonstrative, objectifying I-It (*Ich-Es*). Elaborations on how the loss of things affects this side of our existence would be relevant.

The question raised by the reference to Buber can be taken still a step deeper. It might be that the encounter with another human being's person is the foundational origin of ethics in general. This is Emmanuel Levinas's (1906–1995) claim in his *Totality and Infinity* (1961). Levinas writes that otherness expressed in the Other, particularly in another person's face, is infinite in the sense that it evades ego's attempts to grasp it conceptually. This is similar to Descartes's God, who is infinite in the sense that he overflows all the concepts that try to represent him. For Descartes the infinite and perfect being is the foundation of his metaphysical system. Somewhat analogously, the face of the Other and the ethical demand put forth from the face is the origin of ethics for Levinas. Further, discourse and truth, and therefore all philosophy, depend on this ethical phenomenon. Fundamentally the message of the Other's face is "don't kill me" or "Thou shall not kill."[126]

The main threat to the Other is the ego's tendency to reduce otherness to "sameness," make it understood in ego's conceptual terms. "Totality" in the name of the book refers to philosophical systems that have perfected "the work of the same" and explain all existents by totalizing principles. Consider Heraclitus's "all is war" or Hegel's dialectical logic.[127]

This brief description does not do full justice to Levinas's intricate account. It does, however, show how he solves the foundational problem of ethics: "the grounding" of ethics is not to be sought in reason, the first person's conscience, nature, or even language itself but in a primordial phenomenon that underlies language, truth, and knowing.[128] Therefore, if it should happen that technological development affects the possibility of encountering the demand of the Other, it would also touch the foundation of ethics. If Borgmann's account is valid, it suggests a danger that life structured by the device paradigm is closing possibilities for the originary ethical phenomenon. How things and people in their otherness enable ethical thinking, nonetheless, is still a line of inquiry in need of further study.

Perhaps surprisingly, however, Levinas himself took a positive view of modern science and technology. In "Heidegger, Gagarin and Us," the only essay where he explicitly addresses the topic, Levinas perceives science and technology to be demythologizing nature from its divine pretentions, which opens a truer humanity for human beings.[129] It is not clear, however, how Levinas would have assessed the more recent communication, information, and network technologies. In what sense does one encounter the Other online?

From another angle, various positions that draw on Hegel's philosophy, for instance Taylor's view in his *Ethics of Authenticity*, in anthropology emphasize the human need for recognition (*Anerkennung*) from another human consciousness.[130] The concept of recognition refers to a dialectic of two (or more) self-conscious beings who strive to become acknowledged by the other as conscious subjects, not mere things or instruments. Love, friendship, and respect are typical recognition relationships. It is reasonable to think that the meaning for human life could come—increasingly—from recognition relationships as the material world engages humans less.[131] Borgmann's rejoinder would probably emphasize that without things, places, and shared practices the relationships of recognition become shallower or even impossible. If three people watch TV, for instance, the recognition relationships between them are thin or nonexistent. But if they play a board game—a non-device-structured entertaining pastime—they enter into a playful web of recognition relationships.

Borgmann's Concept of Technology

Borgmann has been accused of an essentialist conception of technology. For example, Andrew Feenberg and Larry Hickman have noted that Borgmann's way of interpreting the technological development in modern societies does not do justice to the plurality of technological phenomena.[132] It appears that Borgmann does not take into account the vast body of science, technology, and society (STS) studies, since if he did, he could not hold such a uniform view of technology.[133] Feenberg also remarks that Borgmann is reluctant to notice how actively and creatively people receive technological devices. For instance, the Internet and its predecessors have made new kinds of communal life possible, and the users have made good use of this potential.[134]

This criticism of the monolithic conception of technology is relevant but not fatal to Borgmann's position. He has succeeded in creating a compact theory that provides a deep understanding of the contemporary culture. Furthermore, his approach analyzes insightfully technology as a way to be in the world. This is something that mere case studies cannot provide.

Defining and Testing Focal Practices

Borgmann's treatment of his central concepts leaves room for the question: What exactly is a focal thing? One cannot avoid feeling that Borgmann's examples are somewhat romantic and old fashioned. Taking care of a wood stove is a focal practice, while computer gaming certainly is not. How much personal taste or idiosyncrasy is involved in Borgmann's account of focal things and practices? Is Borgmann nostalgic?

Some nostalgia, however, can be forgiven, since nostalgia is in harmony with Borgmann's general argument. If the device paradigm is the throughgoing pattern of technology, then it is evident that things free from this technological pattern are found in the past, since no one probably would deny the ever-increasing role of technology in our culture. Yet I still think his account remains open for accusations of backwardness and romanticism.

Larry Hickman has proposed that focal practices and their wholesome effects should be empirically tested.[135] Borgmann has rejected this possibility. According to Borgmann, we can fail in our evaluations of focal things and practices, but in principle they are not objectively testable.[136] I find Borgmann's position understandable. Testing focal things would mean some kind of operationalization, or even quantification, which would contradict the context-related nature of focal things and their indefinite aspects. However, inability to evaluate focal things without ambiguities means that questions concerning them remain open.

Focal practices as they are described in this inquiry are vulnerable to yet another general criticism. An uncharitable reader could ask: "Is he saying that 'good hobbies' will save us from the problems of technological modernity?" In fact, as was shown earlier, Borgmann does have substantial ideas considering larger socioeconomic settings, and these are related to the role of focal things in our way of life.[137] Therefore, the uncharitable question somewhat misses the point. Still, if one considers, for instance, farming as a focal practice in contrast with running and fly-fishing, it is evident that farming is a comprehensive way of life and livelihood, which the others are not: they are hobbies.[138] The problem is that farming as a livelihood is no longer available for—and probably not even desired by—most people.[139] The same goes for other traditional occupations such as those of artisan or even homemaker.

More analytically put, it appears that two quite different conceptions of engagement are at work in Borgmann's thinking about things and practices.[140] First there is the pretechnological engagement, heating the cottage to survive a tough winter, for instance. Then there is the "post-technological" voluntary burden that runners, fly-fishers, and home chefs carry.

Furthermore, if from one angle, focal things and practices look like no more than serious hobbies, from another angle, they appear as cumbersome, laborious, and artificial efforts. Traditionalism involved, for instance, in the culture of the table can be perceived as suffocating stiffness or empty formalism as well as invigorating focal situation. At least partly, the simplification of cultural patterns during the modernization process appears to have been motivated by an emancipatory desire to jettison forms that reinforced inequality and obstructed human freedom as well as authenticity. Paradoxically, in the end of this development, Borgmann—along with others[141]—is arguing for re-instituting some of the old forms. Perhaps people can find new meaning in efforts and forms now that they have the option of being free from them. Yet it often appears that focal situations pose a challenge that threatens to lead one into pretentious self-consciousness rather than authenticity.

An argument to "save" Borgmann here might go along the lines of Charles Taylor's *Ethics of Authenticity* (1992). Taylor discusses the dialectical relationship of authenticity and external demands:

> Otherwise put, I can define my identity only against the background of things that matter. But to bracket out history, nature, society, the demands of solidarity, everything but what I find in myself, would be to eliminate all candidates for what matters. Only if I exist in a world in which history, or the demands of nature, or the needs of my fellow human beings, or the duties of citizenship, or the call of God, or something else of this order matters crucially, can I define an identity for myself that is not trivial. Authenticity is not the enemy of demands that emanate from beyond the self; it supposes such demands.[142]

His point appears to be that authenticity is not possible without a dialogue with some requirements—needs, necessities, duties, ideals, or standards, for instance—coming from outside, since merely self-referring authenticity leads to arbitrary subjectivism and/or tyranny of instrumental reason. Hence, from Borgmann's viewpoint, an analogous argument could state that personal existence cannot truly be deep and authentic without a dialogue with things and practices—that stand out in firm and commanding otherness.

One last question related to things and practices concerns the deictic discourse within which one speaks about them. Borgmann's description of it is, no doubt, analytically relevant. Yet it appears that some further steps concerning it are needed. In deictic discourse we are not able to prove, argue, or practice critique in the same sense as in apodeictic and paradeictic discourses. One could say that in these two, calculative or critical thinking is characteristic, but in deictic discourse meditative or reflective thinking reigns. This distinction seems relevant when someone speaks about art, wilderness, or uplifting practices, but the languages of advertisements or fanatic religious sermons and political speeches are deictic as well—this is fair to accept. So, how can one be critical within the deictic discourse? How can one judge the claims if critical thinking is somehow bracketed in the discourse? These questions still need further elaboration. They, in fact, relate to the theme of thinking about the condition of the good life without giving the good life too "thick" or distinct a content. Borgmann wants the good life to be steered by things more than by ideology. But there must be some kind of dialectic between "listening to the things" and thinking.[143]

Focal Things and Practices in Japan[144]

One way of assessing focal things and practices, different from but in the spirit of Hickman's proposal, would be by cross-cultural comparison. If things like what Borgmann calls focal things and practices could be found in other cultural contexts, then this would be a strong support for his argument. Indeed, I would suggest that sports, arts, pilgrimages and other religious activities, sauna culture, storytelling, and more are examples of focal things and practices in many cultures. The difference between earlier things and practices and their contemporary equivalents is that the focal nature of earlier practices was seldom consciously treasured. The mother who kept fire in the hearth hardly noticed any special satisfaction or depth in her work; because of this, she would often have been happy to give up her burden.

There are, however, a few interesting examples of consciously retained focal practices in diverse cultures, comparison with which may bring a new appreciation for Borgmann's analyses. Perhaps one of the best such examples is to be found in Japan, where traditional culture exhibits various practices with explicitly focal characters that have indeed been highlighted by the encounter with modern technology.[145]

Consider the case of Kendo, an art-of-sword practice the roots of which go back to prehistoric Japan, but which was formed into a conscious focal practice with elaborate philosophy during the Tokugawa Period (1600–1868). Today the tradition of Kendo is preserved by the All Japan Kendo Federation (founded 1952). According to *The Concept of Kendo*, the official statement of the Kendo Federation, "[t]he concept of Kendo is to discipline the human character through the application of the principles of the Katana (sword)." Later, the text speaks of cultivating a vigorous spirit, associating with others with sincerity, and forever pursuing the cultivation of oneself. Is this not exactly the kind of centering, exertion, and striving for excellence for which Borgmann is looking?[146]

Another example of a Japanese focal practice is the tea ceremony, Chado. The idea behind the tea ceremony is to make, share, and drink tea observing the principles of harmony, respect, purity, and tranquility. In short, the tea ceremony is a focal practice created around the focal thing, tea. Kendo, the Way of the Sword, and Chado, the Way of Tea, present focal thing and focal practice structures even in their names. The aim of cherishing focal things within focal practices is also evident in the arts of gardening, flower arrangement, calligraphy, and brush painting. In modern times such new practices as Judo and Aikido have been formed from classical martial arts. The idea of all these cultural practices is to come to know oneself, others, and reality more deeply through practicing the respective art.[147]

In general, the Japanese culture values material things, especially carefully crafted ones but other things as well, more than the Western. This is also manifested in language. For instance, chopsticks, *hashi*, receive the honorary prefix "o," and hence they are usually called *o-hashi*.[148] The same applies to tea, which is called *o-cha*.

The Japanese "focal" practices owe their philosophical and spiritual ideas to various sources. Evidently, there is a Shinto influence, as well as Confucian, Taoist, and Buddhist ingredients in the concepts and practices of these Ways.[149] Here I will pay a little more attention to the Zen Buddhist ideas that have affected swordsmanship and the tea ceremony.[150] These practices, however, may appear as elitist phenomena or bound to the class society of feudal Japan. This doubt, in fact, contains a relevant point, since it hints at a deeper understanding of focal practices. In Zen Buddhism, even quotidian concrete tasks, such as cooking, chopping wood, and carrying water, can be seen as spiritual practices when they are performed with free mind and fully aware. Thus disburdenment from them would mean a spiritual loss for a Zen practitioner. In fact, nearly anything can become a focal practice for a Zen practitioner. The founder of the Japanese Soto school of Zen, Dogen Zenji (1200–1253), wrote the text "Instructions for the Cook" (*Tenzo Kyōkun*), in which he implies that the cook's office in a monastery provides an excellent opportunity to practice Zen, better than reading or chanting sutras. In Dogen's teaching, undivided activity in ordinary tasks leads to insight and realization that transcend all the conceptual dualistic oppositions. Of these antagonisms, existentially the most pressing ones are the contrapositions of life and death as well as ego and non-ego. Moreover, Dogen emphasizes overcoming likes and dislikes.[151]

The argument here, then, is that Japanese practices such as Kendo, Judo, and Chado are essentially focal practices involving focal things in the sense that Borgmann applies these concepts. Since Japan has long and philosophically reflected traditions of focal practices, it is clear that, in general, Borgmann's argument is not as idiosyncratic or unrealistic as it might appear at times. Acknowledging this parallel provides support for Borgmann's stance that consciously cherished focal things and practices may serve as a ground for a deep and meaningful life within the technological world. Without them, life lacks gravity and loses reality. Furthermore, focal things provide a metatechnological viewpoint that enables us to weigh technological devices.

CONCLUSION

This chapter has focused on the life ethics in a technological world. The discussion sprang from the question whether technological advance has made citizens' lives better in technologized societies. In chapter two, this question was placed under the rubric "Challenges for Ethics." Yet while studying Borgmann's thinking it became evident that reflection on the question of the good life raises a number of questions related to applicability and foundation issues. Although various relevant criticisms of Borgmann's work need to be discussed, I believe that this chapter provides some positive conclusions.

First of all, it is evident that technology as material devices is far from neutral concerning the good life. This non-neutrality is manifested in the way in which the technological artifacts structure the human lifeworld by modifying personal relationships, altering the relation between means and ends, between body and mind, between individual and community, for instance. Moreover, the new artifacts simply occupy space and time in everyday surroundings and require attention. This space, time, and attention given to the new technologies means taking them away from other human activities and things related to them. Second, the above point evades conscious reflection in our culture: it is not discussed, or technology is perceived to a large extent as morally neutral, or it is taken for granted that new technologies provide new possibilities that provide a better life. Due to this lack of conscious reflection on morals in technology, the moral effect of technological advances is mostly implicit, silent, unnoticed, and unconscious. The devices settle many questions humans never asked. Third, ethics consequently should pay more attention to the material culture: things, devices, practices, places, institutions, and policies. All these mold the contexts of moral agents. In fact, they do more: they mold the moral agents themselves.

One of the main problems in Borgmann's approach is that for many it is not well enough informed by empirical research on the phenomena it addresses. Larry Hickman is one of Borgmann's critics who argue along these lines. The next chapter will introduce Hickman's own position in more detail.

NOTES

1. Alfred Nordmann, in "Technology Naturalized: A Challenge to Design for the Human Scale" (*Philosophy and Design: From Engineering to Architecture*, 2008), develops this observation further.
2. Albert Borgmann, "Finding Philosophy," *Falling in Love with Wisdom: American Philosophers Talk about Their Calling* (1993), 157.
3. Ibid., 159.
4. Borgmann, "Finding Philosophy," 158–159; *Technology and the Character of Contemporary Life: A Philosophical Inquiry* (1984, hereafter *TCCL*), vii, 15–22.
5. David Strong and Eric Higgs, "Albert Borgmann's Philosophy of Technology," *Technology and the Good Life?* (2000), 25–26. Borgmann especially shows his favor for an older colleague from the University of Montana, Henry Bugbee, the author of *Inward Morning* (1958), and Missoulian author Norman MacLean, best known for *A River Runs through It* (1976). Bruce Wilshire's *The Primal Roots of American Philosophy: Pragmatism, Phenomenology, and the Native American Thought* (2000) offers a compact description of Bugbee's philosophy and a general overview of the roots of American philosophy.
6. Carl Mitcham, *Thinking through Technology: The Path between Engineering and Philosophy* (1994), 19–61.
7. Ibid., 13–14, 17, 29–31, 38–39.
8. Ibid., 13–14, 17, 39–61, 62–78.
9. "To say that the world is real and intelligible is to say that it is lawful. [. . .] From this it follows, as both Aristotle and Kant argue, that to render something intelligible is to place it explicitly in the matrix of laws and principles" (Borgmann, *TCCL*, 19).
10. Borgmann, *TCCL*, 18–22.
11. There are a few commonly applied meanings for the words "instrumentalism" and "instrumentalist." In philosophy of science (e.g., Mary Hesse, "Laws and Theories," *The Encyclopedia of Philosophy*, 1967) they refer to a view that perceives scientific theories and concepts as instruments, tools, or devices. These instruments enable control, prediction, and manipulation of the situations that they are applied to. Thus instrumentalism is a contrasting view to scientific realism that perceives concepts and theories as primarily describing reality. In the philosophy of technology instrumentalism, however, has a different meaning.
12. Borgmann, *TCCL*, 15–17, 26–29.
13. Borgmann, *TCCL*, 15–17, 22–26; *Real American Ethics: Taking Responsibility for Our Country* (2006), 105.
14. Borgmann, *TCCL*, 22–26. C. P. Snow, *The Two Cultures* (1993, first edition 1960).
15. Borgmann, *TCCL*, 9. Larry Hickman, *John Dewey's Pragmatic Technology* (1990), 8–10.
16. Borgmann, *TCCL*, 9–10. The previous chapter showed how technology remained opaque in Hans Jonas's treatment.
17. Borgmann, *TCCL*, 10–11. In public policy one most probably finds an instrumental understanding of technology. See also Langdon Winner, *Autonomous Technology: Technics-out-of-Control as a Theme in Political Thought* (1977), 226–236; Hickman, *John Dewey's Pragmatic Technology*, 8–14.

18. Borgmann, *TCCL*, 11. *Technology and the Character of Contemporary Life*, which is still Borgmann's central work focusing on technology, was written in the early 1980s. Then, the social-constructivist movement in technology studies was only emerging. What Borgmann calls a pluralist view in his 1984 book has since become a predominant approach in academic technology studies. A concise and accessible example of this type of research is Harry Collins and Trevor Pinch's *The Golem at Large: What You Should Know about Technology* (1998). Wiebe Bijker's *Of Bicycles, Bakelites, and Bulbs: Toward a Theory of Socio-Technical Change* (1995) is a more research-based presentation of constructivism. Bijker's "Understanding Technological Culture through a Constructivist View of Science, Technology, and Society" (*Visions of STS: Counterpoints in Science, Technology, and Society Studies*, 2001) is a brief characterization of the constructivist position. Andrew Feenberg, in *Questioning Technology* (1999), has aimed to take constructivism into account without losing the edge of his critical theory of technology. Borgmann has not discussed much social constructivism in his later books. His view on it, however, seems to be rejective. This obviously fits well with his scientific realism. For instance, Collins and Pinch's earlier book *Golem: What You Should Know about Science* (1998, first edition 1993) provides a rather different view of science than Borgmann's realism. For Pinch and Collins science is a characteristically socially constructed—and conditioned—*craft*, not a realistic description of lawful reality. Bruno Latour's *We Have Never Been Modern* (1993) presents a comprehensive philosophical account of constructivist understanding of science and society.

19. Borgmann, *TCCL*, 11–12.

20. All these terms are derived from the Greek verb *δείκνυμι*, which has the range of meanings "bring to light," "show forth," "portray," "present," "render," "show," "point out," and "set before one." The Greek derivative *δεῖγμα* has the meanings "sample," "pattern," "plan," and "sketch." George Henry Liddell and Robert Scott, *A Greek-English Lexicon* (1983).

21. Borgmann, *TCCL*, 71–72, 176–182. Strong and Higgs, "Albert Borgmann's Philosophy of Technology," 36–37.

22. Borgmann, *TCCL*, 176–177.

23. Borgmann, *TCCL*, 15–21, 71–72. Strong and Higgs, "Albert Borgmann's Philosophy of Technology," 36–37. Borgmann uses the word "apodeictic" in an unusual manner. In his use it means quite simply science as it is today practiced. Hence he does not refer to any specially certain form of knowledge.

24. Borgmann, *TCCL*, 71–72, 179–182. David Strong, "Philosophy in the Service of Things," *Technology and the Good Life?* (2000), 325–329. Strong and Higgs, "Albert Borgmann's Philosophy of Technology," 36–37. The point that science is unable to articulate the priorities in reality and thus unable to assist humans in orientation in life is nuanced modification of the theme that science undermines ethics. "Orientation" obviously does not yet concern ethics, but the possibility to discuss orientation is a necessary condition for ethical discourse.

25. Borgmann, *TCCL*, 73 (emphasis mine). Borgmann's word choice, "explanation," is quite strong here. Borgmann acknowledges his debt to Thomas Kuhn (*The Structure of Scientific Revolutions*, 1970, 43–51) concerning the concept 'paradigm,' but also says that his use of the concept is different from Kuhn's to the extent that this relationship is not really relevant to discuss here.

26. Borgmann (*TCCL*, 75) is critical of the methodology in social sciences (according to him, the mainstream in the United States) that aims at value-free law-like precision and predictive power. This kind of methodology is obviously trying to follow the example of the apodeictic explanation of natural sciences.

27. Borgmann, *TCCL*, 25–26, 68–78, 179–182.

28. Borgmann, *TCCL*, 64. On the contrary, Frederick Ferré in his introductory book *Philosophy of Technology* (1995, 11, 87ff) gives an impression that "axiology of technology" can be created just by applying axiology to technology.

29. Albert Borgmann, *Power Failure: Christianity in the Culture of Technology* (2003, hereafter *PF*), 17–18, 31.

30. Borgmann, *TCCL*, 76.

31. For instance, in 1868, due to the unusually severe winter, approximately eight percent of the Finnish population died from hunger and the diseases following it.

32. Albert Borgmann, *Crossing the Postmodern Divide* (1992, hereafter *CPD*), 20–27. Francis Bacon: *The New Organon and Related Writings* (1960), e.g., I, 1–3, 73, 74, 81, 121, 124, 129. Also René Descartes, e.g., in *Discourse on Method,* (2007, 48–57) is explicit about the technical power of science.

33. Borgmann, *TCCL*, 35–40; *CPD*, 20–27.

34. Borgmann, *TCCL*, 37–38.

35. On the relationship of science and technology see, e.g., Mitcham, *Talking through Technology*, 199–204. Borgmann himself might not exactly follow the lines of this characterization. It is, however, possible to maintain conceptual category distinction (that Borgmann perceives important) between science and technology, even if practically, socially, and historically they are intertwined in the above manner.

36. Borgmann, *TCCL*, 85–101; *CPD*, 81; *Real American Ethics* (hereafter *RAE*), 143.

37. Borgmann, *TCCL*, 36–40, 126, 130–135, 142–143; *RAE*, 62–65.

38. Borgmann, *TCCL*, 126. In his 1984 book Borgmann's use of the ideal of excellence appears as much heuristic as substantial. The heuristic purpose is to highlight the features of contemporary life. In *Real American Ethics* (99–138) he constructs explicit substantial ethics along similar lines.

39. Borgmann, *TCCL*, 124–143; *RAE*, 94–98.

40. Borgmann, *TCCL*, 126, 130–131; *CPD*, 20–27; *RAE*, 158–160.

41. "One finds that during the twenty-five years after the Second World War, while the standard of living more than doubled, professed happiness declined rather steadily and significantly, and one finds that people in the technologically advanced countries are no happier than those in less-developed ones" (Borgmann, *TCCL*, 130). Later social scientific research, however, has proved this conclusion incorrect. Rather it appears that people in rich societies profess more happiness than those living in poor conditions. Gordon Brittan, "Technology and Nostalgia"; Borgmann, "Reply to My Critics," *Technology and Good Life?* (2000), 78–81, 357–358. Brittan refers to Ruut Veenhoven's *Conditions of Happiness* (1984), a survey of 245 empirical studies on happiness, in his critique of Borgmann's conclusion. Borgmann (*RAE*, 62–67) continues the treatment of the topic and reviews the Day Reconstruction Method (DRM) for measuring happiness in individuals. DRM aims at reducing "the errors and biases of recall" in happiness self-perception (*RAE*, 64).

42. Borgmann, *TCCL*, 140–143. On television: *RAE*, 64, 66, 114–116.

43. Borgmann, *TCCL*, 33–34, 40–41.

44. Borgmann, *TCCL*, 41–42. Nathaniel Hawthorne's short story, or rather essay, "Fire-Worship" (*Tales and Sketches*, 1996, 841–848) elaborates on the same phenomena as Borgmann's analysis.

45. Borgmann, *TCCL*, 42–43.

46. Borgmann, *TCCL*, 40–48, 76–77; *CPD*, 82–97; *PF*, 16–19.

47. This example is my own, not from Borgmann's text.

48. For reflections on compact discs see Albert Borgmann, *Holding On to Reality: The Nature of Information at the Turn of the Millennium* (1999, hereafter *HOR*), 20, 195, 120, 181–182.

49. Borgmann, *TCCL*, 43–44.

50. Borgmann, *TCCL*, 123.

51. Thomas Michael Power discusses, in "Trapped in Consumption: Modern Social Structure and Entrenchment of the Device" (*Technology and the Good Life?* 2000, 279–281), economics and the device paradigm. Economic theory uses concepts like the market mechanism and the economical system. Combining these ideas with the ongoing process of global liberation of economy one could say, at least metaphorically, that the biggest device today is the globe itself. Incidentally, it has already been referred to long ago as Spaceship Earth by R. Buckminster Fuller (see Borgmann, *TCCL*, 146–147).

52. Pekka Himanen, in *The Hacker Ethic: A Radical Approach to the Philosophy of Business* (2001), claims, among other things, that the transition from the industrial age to the information age enables people to find a new, more creative and satisfying relationship to work.

In Himanen's model, work is an end itself, not mere means to earn money for having fun—work is fun. It still remains to be seen how widely this kind of "hacker ethic" could spread within the present economic structures.

53. An account of technologically structured work: Albert Borgmann, "Society in the Postmodern Era," *The Washington Quarterly* (2000); *TCCL*, 114–124; *CPD*, 16–18, 45, 70, 75, 85, 101.

54. Borgmann, *TCCL*, 48–56.

55. Borgmann, *TCCL*, 125, 139. Here, one could protest against Borgmann's claim that the foreground of technology requires no skills or effort from the user. Cell phones, stereo systems, and video recorders, not to mention computer software, are not that easy to use, at least if one wishes to take advantage of all the features.

56. Borgmann, *TCCL*, 140. In modern literature Aldous Huxley's *Brave New World* (1932) presents one of the most insightful and prophetical accounts of entertainment. This is one reason why it still is a startling vision of a dystopia, and in this sense far more frightening than George Orwell's *Nineteen Eighty-Four* (1948). Although the developments in the early 2000s in connection to the "War on Terror" made Orwell's nightmares relevant again, the soft tyranny of entertainment still appears a more central phenomenon in advanced societies.

57. Borgmann, *TCCL*, 41–42.

58. Borgmann, *CPD*, 114–116, 141.

59. Borgmann, *PF*, 19.

60. Borgmann (*CPD*, 82–83) says he has adopted the term "hyperreality" from Umberto Eco (1986) and Jean Baudrillard. He (*HOR*, 184) says that sometimes it is synonymous with virtual reality.

61. Borgmann, *CPD*, 82–97; *HOR*, 181–192; *PF*, 28–29, 125.

62. Borgmann, *CPD*, 93–94; *PF*, 15–16, 22.

63. Borgmann, *TCCL*, 78–80.

64. In Heidegger's analysis of "nihilism" ("The Word of Nietzsche," *The Question Concerning Technology and Other Essays*, 1977, 53–112), the talk about values indicates that the culture is already in nihilistic condition. If one can choose one's values, the true prescriptive meanings have been lost. "No one dies for mere values," Heidegger (142) remarks. See also Hubert Dreyfus, "Heidegger on the Connection between Nihilism, Art, Technology, and Politics," *The Cambridge Companion to Heidegger* (1993). Borgmann's analysis echoes these Heideggerian themes in a more cautious manner.

65. Borgmann, *TCCL*, 80.

66. Borgmann, *TCCL*, 78–85. See also Langdon Winner, "Brandy, Cigars, and Human Values," *The Whale and the Reactor: A Search for Limits in an Age of High Technology* (1986), 155–163.

67. Borgmann, *TCCL*, 78–80; *CPD*, 110–111.

68. Borgmann, *TCCL*, 81–85. In passing, Jacques Ellul (Mitcham, *Thinking through Technology*, 57–58) repeatedly argued that if Marx himself had written in the 20th century, he would have focused on technology instead of economic relations. Ellul perceived his own sociological project as work on this task.

69. Charles Taylor's *Ethics of Authenticity* (1992) in general addresses the problems of moral subjectivism and instrumental reason in liberal democracies. In the chapter "The Slide to Subjectivism" (55–69) he addresses the particular question of the privatization of the good, and one of the conclusions is the following: "This [self-determining freedom] sets up a vicious circle that heads us towards a point where our major remaining value is choice itself" (69). In the philosophical tradition Hegel's treatment of Good (*das Gute*) in *The Elements of the Philosophy of Right* (§§ 129–132) offers an elaborate criticism of solitary valuations.

70. Albert Borgmann, "The Moral Significance of the Material Culture," *Technology and the Politics of Knowledge* (1995); "Material Culture," *Encyclopedia of Science, Technology, and Ethics* (2005).

71. Borgmann, *TCCL*, 101 (emphasis mine). Sami Hautakangas and Tomi Kiilakoski, in "The Information Society: Towards an Iron Cage of E-learning?" *European Educational Research Journal* (2004), analyze and criticize e-learning in a Heideggerian vein that comes close to Borgmann's account of the device paradigm. Borgmann's thinking, in fact, is highly useful in assessing education and technological solutions offered for enhancing education.

72. Langdon Winner, in his influential article "Do Artifacts Have Politics?" (*The Whale and the Reactor*, 1986, 19–39, first published 1980), makes a distinction between two different ways technological artifacts can have political characteristics. (Obviously, having politics and being good or bad are different questions, but in this context they are somewhat closely related.) First, he notes that technological artifacts can be disguised ways to practice politics, i.e., technology is the means for political ends. For instance, the application of different transportation systems can be a means to advance political goals. Second, the technological solutions can be in a more inherent way political. For instance, nuclear power appears to require certain societal arrangements that secure the safety in the plants, during the transportation of waste, and so on. These arrangements, at least it can be argued, cannot be fully democratic in nature. So, in this sense, nuclear power is "inherently political."

73. Borgmann, *CPD*, 110 (underlining mine).

74. Borgmann, *TCCL*, 209–210; *PF*, 22–24.

75. Borgmann, *TCCL*, 196–197.

76. Borgmann, *TCCL*, 197.

77. Borgmann, *TCCL*, 199. Borgmann is referring especially to the text "Turning" (see Heidegger, *Question Concerning Technology*, 43).

78. Heidegger's text "The Thing" (*Poetry, Language, Thought*, 2001) is in the background of Borgmann's thinking on things.

79. Borgmann, *TCCL*, 80–81, 159, 191–192. Strong, "Philosophy in the Service of Things," *Technology and the Good Life?* (2000), 327–329. Science does not treat beings as things either. Science is crucially dependent on its ability to create research objects, conceptions that appear in a similar manner to all observers that are measurable—preferably in clear quantities. Hickman (*John Dewey's Pragmatic Technology*, 86–89) summarizes John Dewey's analysis of objectification. It has a thematic link to Heidegger's and Borgmann's considerations. Dewey presented it in connection to his depiction of Australian aborigine culture, in which objectification does not exist. According to Dewey, four aspects characterize objectification: (1) objects are understood as *constant* and therefore possible to characterize mathematically, (2) objects can be *substituted*, (3) objects can be *controlled*, and (4) they can be understood within a matrix of scientific *laws*.

80. Borgmann, *TCCL*, 80–81.

81. Borgmann, *TCCL*, 44–47.

82. See, e.g., Maclean, *A River Runs through It and Other Stories* (1976), Hemingway, "Big Two-Hearted River," *The Complete Short Stories of Ernest Hemingway* (1987).

83. Borgmann, *TCCL*, 201–204. Henry David Thoreau, "Walking," *The Portable Thoreau* (1964), 592–630.

84. Borgmann, *TCCL*, 204–206.

85. Ibid., 202.

86. Ibid., 196–197.

87. Ibid., 169, 182–196.

88. Ibid., 169.

89. Ibid., 176–177.

90. Borgmann, *CPD*, 116–126. This is one of the many Aristotelian features in Borgmann's thinking: the focal reality is found in the mean.

91. Borgmann, *CPD*, 116–122; *TCCL*, 206–208.

92. *Nicomachean Ethics*, 1098a 15–21.

93. A metaphor that could be used to describe, on the one hand, life within the device paradigm and focal reality, on the other, is that of lightness and weight. Life in hyperreality appears to manifest "the unbearable lightness of being," whereas focal things bring weight and gravity to life. Borgmann ("Heidegger and Ethics beyond the Call of Duty," *Appropriating Heidegger*, 2000, 72) refers to Milan Kundera's novel that introduced the term.

94. Borgmann, *TCCL*, 207.
95. Borgmann, *RAE*, 90–94, 136–138.
96. Borgmann, *TCCL*, 215.
97. Heidegger, *Being and Time*, §§ 27, 35–38.
98. Heidegger, *Being and Time*, §§ 51–63.
99. Borgmann, *TCCL*, 211.
100. Borgmann, *TCCL*, 221.
101. Borgmann, *TCCL*, 215, 247–248. This is the central idea in David Strong's *Crazy Mountains: Learning from the Wilderness to Weigh Technology* (1995), which develops Borgmannian ideas.
102. If the device paradigm in itself is not taken as the underlying unquestioned good of the modern society.
103. Borgmann, *TCCL*, 211–214.
104. John Rawls, *Theory of Justice* (1971, 426). Borgmann, *TCCL*, 213.
105. Borgmann, *TCCL*, 214.
106. Borgmann, *TCCL*, 213–218. Sometimes focal interests conflict. An example of this was the dispute on the building of a new concert hall in Helsinki in the early 2000s. It appeared as a conflict of interests between the advocates of "folk culture" and the advocates of "high culture." The former wanted to preserve old railroad warehouses—which had to be torn down to make room for the concert hall—in the city center, since they had become an important base for various grassroots and communal activities. They saw the concert hall as an elitist project. The advocates for the hall probably had nothing against the grassroots activities but nonetheless perceived that the city center belongs to "higher" pursuits. Both approaches stressed focal interests but from different perspectives—building an office complex or a shopping mall would have been the device paradigm choice.
107. Borgmann, *TCCL*, 223–224.
108. Borgmann, *TCCL*, 223–224, 234–236. Borgmann (*CPD*, 72–75) discusses how a postindustrial economy starts to some extent to create a need for "flexible specialization" that can be rather free from the pattern of technology.
109. Borgmann, *TCCL*, 235. A European reader may notice that although Borgmann's cultural analysis applies for the most part to all the technologically advanced societies, it is explicitly based on observations concerning the United States. This more often than not means that the commodification and consumerism described by Borgmann have less extreme forms in Europe, and some of his pleas have long been taken into account by many European governments and local communities. For instance, in most European cities the city centers are focal things—or focal places—while in the United States the role of the city center has been taken over by shopping malls, which indeed are manifestations of the device paradigm.
110. Borgmann, *TCCL*, 226–230; *RAE*, 125–138.
111. When Galbraith wrote his *Economics and Public Purpose* in 1973, the numbers were 1,000 corporations and twelve million small firms in the United States. The economy was divided more or less half and half between these forms.
112. Borgmann, *TCCL*, 227, 237ff.
113. Ibid., 220.
114. Strong, "Philosophy in the Service of Things." Borgmann's *Power Failure* and *Real American Ethics* are obviously articulate in religious and ethical content.
115. Borgmann, *TCCL*, 17.
116. Ibid., 20.
117. Heidegger, *What Is Called Thinking?* (1966), 8, 13, 33, 38–47. Edmund Husserl, *The Idea of Phenomenology* (1999), 1–11; *Die Idee der Phänomenologie* (1973), 1–11. Dan Zahavi, *Husserl's Phenomenology* (2003), 72–77, 125–133. Juha Himanka, "Before and After the Reduction," *The Journal of the British Society for Phenomenology* (2001).
118. "Science makes the jug-thing into a non-entity in not permitting things to be the standard for what is real." "Science's knowledge, which is compelling within its own sphere, the sphere of objects, already had annihilated things as things before the atom bomb exploded" (Heidegger, "The Thing," 168).
119. Heidegger, *Being and Time*, § 44.

120. Heidegger, *Discourse on Thinking: A Translation of* Gelassenheit (1966), 43–57.
121. Heidegger, *Discourse on Thinking*, 43–57; *Question Concerning Technology*, 130–136, 176–182.
122. *Nicomachean Ethics*, Book VII.
123. *Nicomachean Ethics*, 1172a–1176a.
124. *Nicomachean Ethics*, Books VII, X.
125. Martin Buber, *I and Thou* (1996, first published in German 1923). Among the classics of our philosophical tradition, Kant's ethics sets a fundamental importance on the respect for a subject's autonomy (*The Grounding for the Metaphysics of Morals*, 1993, 34–44, Ak. 427–440). It appears that a clear and meaningful way to assess technologies is to analyze to what extent they enable or prevent the practice of Kantian autonomy of the will.
126. *Totality and Infinity*, 194–219. I am indebted to David McDonald for insights into Levinas's relevance for ethics of technology. Some of what follows I have learned from him in private conversations and from his unpublished lecture at St. John's College, Santa Fe, NM, on July 27, 2011.
127. Levinas, *Totality and Infinity*, 21–52.
128. Levinas, *Totality and Infinity*, 194–214. Levinas's ideas obviously are not novel. The duty toward the stranger (who is one of the main instances of encountering the Other) and its fundamental role for humanity and/or piety is central in both the Greek and Hebrew traditions. Consider lines from the scene in *The Odyssey* (VI, 206–208), in which Nausikaa receives Odysseus: "But, since this is some poor wanderer who has come to us, / we must now take care of him, since all strangers and wanderers / are sacred in the sight of Zeus, and the gift is a light and a dear one." See also VII, 161–165, 180–81; VIII, 26–29; XXII, 411–417. One way of reading the main plot is that Odysseus's arriving to his house as an unknown stranger and being grossly mistreated by the suitors reveals the suitors' decline below humanity: "These were destroyed by the doom of the gods and their hard actions, / for these men paid no attention at all to any man on earth / who came their way, no matter if he were base or noble" (XXII, 413–415). For stranger, or the Other, in the Hebrew Bible see, e.g., Exodus 23:4–5, Proverbs 3:28–29, and Isaiah 58:6–14. New Testament: e.g., Matthew 7:12, 25:35–46 and Luke 10:25–37.
129. Emmanuel Levinas, "Heidegger, Gagarin and Us," *Difficult Freedom: Essays on Judaism* (1990).
130. In the passage on "lordship and bondage" in *Phenomenology of Spirit*, Hegel (1977, 111–119; 1986, 145–155) introduces the theme that goes through the book. Influential in the philosophy of the 20th century was Alexandre Kojève's (1902–1968) reading of theme (see, e.g., *Introduction to the Reading of Hegel*, 1969, 3–31). More recently, thinkers such as Axel Honneth (*The Struggle for Recognition*, 1996), Charles Taylor (*Hegel*, 1975, 127–170; *Ethics of Authenticity*, 1992, 43–53), and Francis Fukuyama (*The End of History and the Last Man*, 1992, 143–210) have made use of and further developed the notion of recognition. The notion of recognition, nonetheless, emerges as early as Plato's *Republic* (1970, e.g., 375a–375e, 411a–411e, 436a, 439e–441c). In Plato's thought the middle part of human soul is *thymos*, often translated as "spiritedness." Appetite exists in the belly, reason in the head, and *thymos* in the chest. *Thymos* is the part with which humans relate to others: for instance, feel shame and achieve glory.
131. History ends, according to Fukuyama (*The End of History and the Last Man*, 1992), who draws on Kojève's reading of Hegel, in the universal and homogeneous state. In this state liberal democracy, free market, and technoscientific progress coincide. It is the end of history, since history is a drama of emancipation, in which individuals, groups, classes, and nations strive toward freedom and recognition. In the end state, these needs are satisfied when all the citizens are recognized as free persons. Hence no new emancipatory horizon to strive for exists; history in its traditional sense ends.
132. Andrew Feenberg, *Questioning Technology* (1999), 187–188, 191–193; "From Essentialism to Constructivism: Philosophy of Technology at the Crossroads," *Technology and Good Life?* (2000), 298–305. Larry Hickman, "Focaltechnics, Pragmatechnics, and the Reform of Technology," *Technology and Good Life?* (2000), 93ff.

133. Feenberg, *Questioning Technology*, 191–193; "From Essentialism to Constructivism," 299–304.

134. Feenberg, *Questioning Technology*, 191–193; "From Essentialism to Constructivism," 300–304, 308–310. Borgmann (e.g., in *RAE*, 142) has begun to appreciatively recognize the pluralism in people's reception of technologies. This has not, however, altered the main line of his arguments.

135. Hickman, "Focaltechnics, Pragmatechnics, and the Reform of Technology," 98–99.

136. Borgmann, "Reply to My Critics," 356–357.

137. Borgmann, *TCCL*, 157–169, 226–246.

138. Paul T. Thompson, "Farming as Focal Practice," *Technology and the Good Life?* (2000), 167–178. The whole body of Wendell Berry's work argues for agriculture as a central element of the good human life. Hans Jonas (*The Imperative of Responsibility*, 1984, 194–198) discusses hobbies critically. Jonas severely doubts that hobbies could provide a true content and meaning for human life: "Freedom consists and lives in pitting itself against necessity" (198). Therefore, hobbies that are leisure pursuits with contingent character cannot lead to human fulfillment in the deepest sense. "No ideology can conceal from the actors of this scenario [of a "leisure society" that could result from the automatization] the dismal truth that what they are doing does not really matter . . ." (197). For a contrast from an earlier era, consider Hegel's description of system of needs and the estates (*The Elements of the Philosophy of Right*, §§ 189–208) and corporations (§§ 250–256). In Hegel's vision of a rational civil society everyone's work really matters—it reconciles subjective needs with objective necessities. Jonas and Borgmann are looking at conditions in which this kind of integration appears simply impossible.

139. If organic farming and less technologically enhanced forms of agriculture gain wide popularity, this will mean turning to slightly more labor-intensive ways of farming. Still, it will not substantially affect the current trend in the transformation of the occupational structure toward services and information processing.

140. Peter Paul Verbeek, in *What Things Do* (2005, 186–188), analyzes Borgmann's thinking on engagement in this vein.

141. José Ortega y Gasset, in *The Revolt of the Masses* (1950; orig. *La rebelión de las masas*, 1930), was already in the early 20th century troubled by the same question.

142. Taylor, *Ethics of Authenticity* (40–41). Taylor in this 1992 book makes use of Borgmann's ideas from *Technology and the Character of Contemporary Life*.

143. For treatment of the relationship between thinking and things see Strong, "Philosophy in the Service of Things."

144. Some of this section has appeared earlier in Topi Heikkerö, "The Good Life in a Technological World: Focal Things and Practices in the West and in Japan," *Technology in Society* 27 (2005, No. 2), 251–259.

145. Dreyfus ("Heidegger on the Connection between Nihilism, Art, Technology, and Politics," 295, 307) makes remarks on the Japanese culture. They come close to my observations.

146. *The Concept of Kendo* in its entire form and more information about the All Japan Kendo Federation can be found on the website: http://www.kendo.or.jp/ (accessed February 8, 2007). See also Hiroshi Ozawa, *Kendo: The Definitive Guide* (1997) and Minoru Kiyota, *The Shambala Guide to Kendo: Its History, Philosophy, and Spiritual Dimension* (2002).

147. D. T. Suzuki, *Zen Buddhism: Selected Writings of D.T. Suzuki* (1956), 279–294.

148. In passing, the Eastern Asians consider skills required in using *o-hashi* as an essential part of being civilized. Dining with knife and fork is perceived as too easy.

149. D. T. Suzuki, *Zen and Japanese Culture* (1959), 21, 44–45, 48, 57, 91, 304–305.

150. Suzuki, *Zen and Japanese Culture*, 179–294. A classic text on Zen in swordsmanship is Zen Master Takuan Soho's (1573–1645) essay "The Mysterious Record of Immovable Wisdom," which he wrote to shogun's Kendo teacher Yagyū Munenori (1571–1646). Takaun Soho, *The Unfettered Mind: Writings of the Zen Master to the Sword Master* (1986), 19–44. Hans Jonas, in his "Heidegger and Theology" (*The Phenomenon of Life*, 2001)—addressed to protestant theologians—remarks that Heideggerian understanding of Things as gatherings of

the fourfold ill works with Christian theology but brings him to Zen: "At this point [thinking of things] I turn rather to Zen where I find the doctrine of the revelatory and saving function of each and every thing at first hand and on native ground" (251).

151. Eihei Dogen, *Moon in a Dewdrop: Writings of Zen Master Dogen* (1985), 53–68, 124–128.

Chapter Five

Larry Hickman

Technology and Functionality

The cup was strongly made and held a sufficient amount of coffee or tea, whatever you were drinking. The handle was big enough. Even a large man's finger could pass through it. The angle between the bottom and the side was right, so that it was easy to drink from but the liquid didn't splash down your front. The thickness of the material felt pleasant to the lips, and the large shape would cool down a boiling drink quickly, while still preserving its warmth. Someone had thought carefully when designing this cup. Snell said it was Kaj Franck's Teema series from the fifties, famous Finnish design.

Snell became enthusiastic. She took out other dishes, bowls and casseroles from the series, which could be stored one inside another. They carried the stamps of the Arabia factories and the ovenproof symbols. The dishes had been planned both for serving food and for preparing it. Naturally, they could also be used for preserving food. Integrating the whole chain of uses into a single product spoke of surprisingly refined design principles, if it was true that the dishes were from the fifties. Containers and cassettes that could serve unchanged for production, transport and storage did not come into general use in industry until a couple of decades later.

The consul thought the cups were dull and looked monotonous. Heikkilä asked him if the yellow colour didn't nevertheless give them a cheerful look. In the consul's view the colour was crude. He started praising Italian tableware, even the kind in everyday use, which was always refined and a source of aesthetic pleasure to the eye. I had bad experiences of Italian crockery. The coffee cups were so small that you had to purse your lips against the edge as if you were trying to blow a trumpet or some other wind instrument which required the mouth to be precisely tensed against the mouthpiece. When planning a coffee cup, after all, the starting-point should be the function, which is drinking, not ornamentation.

I was so fed up with Italian curves, flourishes and gildings that it felt good to look at the clear lines of Finnish tableware, where the form was determined solely and carefully by the function. Looking at the coffee cup, you could see at once what it was for, and also that it fulfilled its function in the best possible way.[1]

* * *

This scene from Hannu Raittila's novel *Canal Grande* points at the problem of priorities in the design of artifacts and the human world at large. In technical design functionality appears to be one of the guiding principles. But what exactly is functionality? And how big a say should it have in design and in politics concerning technology? Should it perhaps be the governing principle that reveals ethical, aesthetic, or political concerns as irrelevant if they turn out to be disconnected from the function? These are central questions in the pragmatist philosophy of technology that is under scrutiny in this chapter. In it I examine Larry Hickman's Deweyan philosophy of technology. Several reasons speak for taking this position under scrutiny. First of all, in the previous chapters we have noticed that the ethics in technology becomes entangled with various dichotomies of conventional thinking, such as fact/value, means/ends, and individual/community. Now pragmatism claims to recast these problematic dichotomies, which it perceives as results of the foundationalist, representationalist epistemology behind most, if not all, traditional Western thinking. A second reason for paying attention to pragmatism is the recent general interest in pragmatist thought among philosophers of technology.[2] One example of this is Don Ihde's postphenomenology.[3] A third perspective in favor of pragmatism is its alleged role as a bridge builder between the analytic and the phenomenological traditions. This feature of pragmatist thinking is in tune with the basic dialogical commitment of the study, which received support as an adequate approach to the ethics of technology in the second chapter. A fourth point for the relevance of pragmatism is its pronounced emergence in standard philosophy during the last decades of the 20th century. Consider, for example, the approaches of Willard van Orman Quine, Hilary Putnam, Richard Rorty, and Jürgen Habermas.

As I said, Hickman is by no means the only pragmatist philosopher of technology. He, however, represents the classical Deweyan approach to philosophy of technology, while most pragmatically oriented philosophers have developed amalgams of pragmatism and other currents of modern philosophy. Richard Rorty's neopragmatism characterizes the latter approach within the standard philosophy and Ihde's work within the philosophy of technology.

I will start with a brief introduction to pragmatism and its relevance for the philosophy of technology. I will then proceed to describe the central ideas of Larry Hickman's Deweyan philosophy of technology, as he presents them in *John Dewey's Pragmatic Technology* (1990), *Philosophical Tools for Technological Culture* (2001), *Pragmatism as Post-Postmodernism* (2007), and his articles. Hickman's central themes include a recasting of the praxis-theory distinction, anti-foundationalism, anti-essentialism, a broad definition of technology, an emphasis on problem solving and experimentation, knowing as technological artifact, and publics as products. Using these ideas, Hickman develops a perspective that seeks to overcome the limitations of traditional Western philosophy in general and the classical philosophy of technology in particular. Hickman, however, explicitly addresses the ethics of technology only a little. Yet ethics is implicitly present in Hickman's thinking much of the time. Hence the main contribution of this chapter will be an attempt to distill Hickman's implicit ethics and show the central ethical implications of his work. By and large they have to do with reflecting on technological practices, education, and democratic participation. I will conclude with succinct critical remarks on Hickman's project.

PRAGMATISM AND THE PHILOSOPHY OF TECHNOLOGY

American pragmatist philosophy started from a critical stance toward modern representationalist epistemology. Charles Sanders Peirce (1839–1914) presented an interpretation of beliefs as habits of action rather than representations of reality. The classical pragmatists, in addition to Peirce, such as William James (1842–1910), as well as John Dewey (1859–1952) and George Herbert Mead (1863–1931) one generation later, established the pragmatist positions in modern philosophy.

The birthplace of pragmatism[4] was the Metaphysical Club at Harvard University in 1871–1874. The club met by turns in James's and Peirce's offices. Its members came from different disciplines and their intellectual backgrounds varied greatly: Darwin's evolutionary theory, radical empiricism, nominalism, Mill's utilitarianism, Bentham's moral philosophy, American transcendentalism, German idealism, mathematical logic, and Herbert Spencer's general evolutionary theory, for instance, were ingredients in the members' theoretical influences. In addition to these, unitarian natural theology was strongly present at 19th-century Harvard.

The club's discussions bore explicit fruit a few years later, when Peirce published a series of articles in *Popular Science Monthly*. In these essays he articulated some of the grounding stances of pragmatism. In "The Fixation of Belief" (1877), Peirce argued that human beliefs are formed in a process: a belief leads to formation of a habit, humans hold on to the habit until action according to it leads to an unwelcome surprise, surprise leads to doubt, doubt leads to inquiry, and inquiry may lead to formation of a new, more suitable belief that serves as a ground for a new habit, etc. In inquiry, Peirce argued, the self-correcting scientific method works better than the methods of tenacity, authority, or *a priori* speculation. In the scientific method Peirce emphasized its openness to "external permanency" that is not dependent on opinions and is able to affect humans through their senses.[5] In "How to Make Our Ideas Clear" (1878)[6] Peirce brought forth a notion of meaning, according to which the meaning of concepts is in the practical consequences of their referents: "Consider what effects, which might conceivably have practical bearings, we conceive the object of our conception to have. Then, our conception of these effects is the whole of our conception of the object."[7] As generalized, this leads to a definition of reality: "reality, like every other quality, consists in the peculiar sensible effects which things partaking of it produce."[8] Pragmatism for Peirce, however, was not primarily a new philosophy, metaphysic, or even a theory at all, but a method, a technique for solving philosophical and scientific problems.[9]

Although Peirce was arguably the sharpest and most original thinker of the classical pragmatists, it was William James who made pragmatism a well-known current in philosophy. He launched this development in his lecture "Philosophical Conceptions" (1898) and elaborated on his philosophical ideas, for example, in *Pragmatism: A New Name for Some Old Ways of Thinking* (1907). The name of the book refers to James's notion that pragmatism is a method for solving philosophical problems already used by Socrates, Aristotle, Locke, Berkeley, and Hume. This idea is not very far from Peirce's emphases. James, nevertheless, gave content to the term "pragmatism" different in many respects from Peirce's ideas. While Peirce proceeded through a logical phenomenology of human belief-formation, James emphasized more practical, moral dimensions of pragmatism:

> The whole function of philosophy ought to be to find out what definite difference it will make to you and me, at definite instants of our life, if this world-formula or that world-formula be the true one.[10]

John Dewey, who was two decades younger than Peirce and James, became the leading philosopher in North America when James died in 1910. In passing, Dewey's life spanned from the Civil War through the World Wars until the first hydrogen bomb. He thus witnessed firsthand the major social and technological transformations of the modern world. Dewey came from Hegelian and Darwinian backgrounds to appropriate the pragmatist thought. Peirce was Dewey's teacher at Johns Hopkins—but had an impact on Dewey primarily only later on, and Dewey granted that James's thinking had influenced him as well. In Dewey's view, Peirce was a logician, "Kantian" in his manner of setting questions, and a realist, whereas James was a humanist, empiricist, and nominalist. Dewey sought to overcome these oppositions of his predecessors and worked out a general philosophy he called instrumentalism. Moreover, he argued that his new philosophy transcended and recast various modern dualistic dichotomies, of which he found the separation of (scientific) knowledge and morals the most noxious. Consequently, Dewey's instrumentalism aimed to be both a logical theory of science and an approach to ethics and social critique.[11]

At the center of Dewey's instrumentalism is his theory of inquiry. Through inquiry human beings achieve more appropriate ways to exist in their environments and ways to mold an environment to improve their living conditions. Although according to Dewey modern technoscientific research is a paradigmatic form of inquiry, his notion of inquiry is very generally human. Dewey separates four phases in the process of inquiry. It begins with the problematic situation: a state of affairs in which habitual ways of behaving cease to be functional. Problems in Dewey's sense are not narrowly intellectual but always also practical and existential. Second comes the stage of the formulation of a question. It sets the boundaries for the inquiry and defines the question to be answered. The third is the reflective phase, in which intellectual tools such as ideas and theories are assessed as solutions for the problem. The final phase is testing the solution in action. If the solution proves workable, the reflective inquiry ends and the solution becomes incorporated into the arsenal of habitual ways of action. The solutions, however, possess only the status of being "warrantably assertible" conclusions that may be questioned later. This is Dewey's fallibilism: inquiry does not produce certainty and its results are always open to revision.[12]

Sociologist George Herbert Mead, Dewey's colleague at the University of Michigan and the University of Chicago, shared Dewey's socially progressive agenda. Dewey and Mead perceived the technoscientific inquiry as an able means for solving moral and social problems as well as scientific and technical ones. They were not, however, one-dimensionally approving of all technoscientific development: since they understood science as inseparable from the social context in which it is practiced, they criticized what they saw as social malaises in the background of science—the influence of the corporate interest, for instance.[13]

Dewey considered the method of inquiry as an adequate approach in ethics as well. An ontological ground for this uniformity in method was Dewey's stance that there is no real difference between the "subject" of the experience and its "objects," since both exist on the same continuum of nature. Moreover, in experience no sharp distinction between subject and object can be made. Thus, there is no categorical difference between ethics—that would concern subjects' choices—and technoscientific research that concerns the objects in the world. In epistemology, Dewey, moreover, rejected the distinction between moral and non-moral knowledge; according to him, all knowledge has moral aspects, since knowledge is always an instrument of adaptation and thus aims at some "ends-in-view." This view of ethics meant rejecting all absolute and transcendent sources of morals and submitting moral principles for testing in "the laboratory of life."[14]

Like most philosophical "schools" pragmatism can be understood in degrees of broadness. When understood broadly and thematically, aside from its historical beginnings in the United States, pragmatic features can be found in many approaches that are not nominally or historically pragmatist. Consider the following examples. Martin Heidegger's work, especially in *Being and Time* (1927), emphasizes concrete and practical dimensions of human existence.[15] Wittgenstein's later thinking that examines language in the contexts of language games and forms of life arguably emphasizes praxis.[16] Moreover, Marx and Engels's project that centers around material production, work, and the transformation of political conditions evidently possesses a pragmatic vein.[17]

After its first emergence, pragmatism, later in the mid-20th century, appeared to become overshadowed by the logical-analytic philosophy that landed in the United States with the Central European immigrant philosophers. From the 1960s onward, however, pragmatism has restored its relevance in academic discussions. Such varied thinkers as Willard Van Orman Quine, Donald Davidson, Hilary Putnam, Richard Rorty, Jürgen Habermas, Karl-Otto Apel, and Wilfrid Sellars have put the legacy of the classical pragmatists to use. Pragmatism has become crucial both in positions furthering the frontiers of analytic philosophy and as a ground on which dialogue between analytic and phenomenological traditions has appeared promising.[18]

Since pragmatism emphasizes the practical, active side of human life it offers itself as an obvious candidate for an approach to thinking about technology. In the recent philosophy of technology, Paul T. Durbin, Don Ihde, and Larry Hickman have developed pragmatist approaches. Durbin draws mostly upon Mead's work. He has argued for the "application of experimental method to social problems." According to him, experiential reform should be extended to medicine, education, media, industry, information processing, and nuclear power.[19] Ihde has moved from phenomenology toward pragmatism. Ihde's work has centered around analyzing the role of instruments in human existence and in doing science. In these projects he has begun to pay more and more attention to concrete materiality and relations between the human intentionality and artifacts. This has distanced him from the consciousness-centeredness common in phenomenology. Ihde has articulated his debts to pragmatists, especially to Dewey, and called his current approach "postphenomenology."[20]

While many contemporary thinkers make use of pragmatism in amalgams of different intellectual ingredients, Larry Hickman's work could be called neoclassical pragmatism. Hickman aims to follow a purely pragmatist path in his thinking and explicitly calls his position Deweyan. I perceive slight modesty in Hickman's Deweyanism, since he has created a genuine position of his own in the philosophy of technology. Therefore, in what follows I will concentrate on Hickman's thinking, not on Dewey's exegesis.[21]

HICKMAN'S DEWEYAN PHILOSOPHY OF TECHNOLOGY

Hickman's first comprehensive account of the pragmatist philosophy of technology was *John Dewey's Pragmatic Technology* in 1990. The main achievement of the book was to reveal Dewey's thought as a complete account for the philosophy of technology—through a faithful reading of his works. In 2001 appeared *Philosophical Tools for Technological Culture: Putting Pragmatism to Work*, a collection of Hickman's essays from the decade after the first book. The 2001 book presents Hickman more independently defending, clarifying, and developing further his Deweyan position. *Pragmatism as Post-Postmodernism* (2007) collects newer essays and exerts Hickman's pragmatist reflections on areas of cultural philosophy. Next, I will introduce Hickman's definition of technology, move on to elaborate on what he perceives as necessary revisions of some central philosophical distinctions, and observe the way Hickman tackles the problem of technological change.

Defining and Naturalizing Technology

Defining technology has proven a difficult task. Hickman, nonetheless, builds his philosophy of technology on an explicit definition of technology and a philosophical elaboration on the definition. His definition is broad:

> Technology in its most robust sense, then, involves the *invention, development, and cognitive deployment of tools and other artifacts, brought to bear on raw materials and immediate stock parts, with a view to resolution of perceived problems.* Technology in this sense is what establishes and maintains the stable technical platforms—the habitualized tools, artifacts, and skills—that allow us to function and flourish.[22]

The products of technological activity not only consist of devices and engineered artifacts but are diverse: "Sports skills, universities, political parties, and pi are as much technological artifacts as is a hammer."[23]

In Hickman's sense, technological action, which creates the aforementioned kinds of products, is essentially a reflective pursuit. It takes place in research on processes involving techniques. Hickman draws this point out of the etymology of technology in emphasizing that technology is *logos* concerning *technē.* Inversely, as all research needs to employ some kind of instruments, material or mental, all research is technological as well. The Deweyan word here is "inquiry":

> What Dewey thought significant about inquiry, and what he thought discloses its technological character, is that *every reflective experience is instrumental to further production of meanings, that is, it is technological.*[24]
>
> Dewey's definition of *inquiry* is thus more or less his definition of *technology* in its broadest sense.[25]

Carl Mitcham, however, has put forth a critique concerning the broadness of Hickman's definition. He has asked: If nearly all human activities fall in the category of technology what is the meaning and usefulness of such a category?[26] Hickman has clarified his position in *Technological Tools.* According to him, human activities can be categorized in a fourfold manner (see table 5.1). The typology unfolds as follows: first are those activities that are non-cognitive and do not involve tool use; second are those that are cognitive but without tool use; third, those that involve tool use but are non-cognitive; and fourth, those that involve both tool use and cognition. Thus only the last category is "technological." The third consists of acts that are merely "technical." Hickman, moreover, admits the great importance of the activities of the first category that are immediate—not mediated, without tool use—and habitual—non-reflective, with little cognitive element. He says most of human life falls into this category. The second category, consisting of reflective but non-instrumental activities, is "extremely difficult to characterize"[27] and has a minor scope. This is due to Hickman's broad understanding of tools: most reflective processes include at least some inward and mental tools.[28]

Hickman notes that the four categories chart the phases of human experience. Thus the classification is not meant to be evaluative. Most of human life consists of immediate perceptions and habitual responses to them: desiring, feeling pleasure and aversion, and simple action. Although the second category is small, it is necessary, since not all cognitive action employs instruments. Hickman's examples, decision to escape a wild animal or to pick a flower, include very little tool use. Yet given Hickman's broad understanding of technology one could argue that in deciding to escape, the person employs concepts such as the species of the animal, and that the decision to pick a flower refers to the species as well and maybe to some notions of beauty. This shows that the second category is slightly unclear. Some more felicitous examples, however, could be found: a decision to jump over a

Table 5.1. Hickman's Definition of "Technological" against Other Kinds of Human Activities

non-instrumental	*1. habitual, non-instrumental:* immediate perceptions and habitual responses, e.g., feeling of pleasure and pain, smiling, walking (most of human life)	*2. cognitive, non-instrumental:* simple decisions and acts, e.g., deciding to pick a flower (nearly an empty category)
	3. habitual, instrumental: technical—has become automatic, forms a "platform" for action, e.g., driving a car or work on an assembly line	*4. cognitive, instrumental:* technological—problem solving, creativity; adjusting "platforms"; takes place in, e.g., researcher's, artist's, or engineer's work

puddle or to go around it might be one. The third category consists of actions that have become automatic to the extent that they require little conscious attention: various acts in driving a car, using a computer keyboard, and language in spontaneous speech, for instance. Hickman points out that these kinds of technical activities are necessary as "platforms"; without them people's conscious attention would drown in details.[29] Platforms are thus patterns of behavior, building blocks of meaningful living, repetitive actions that provide structure for life. The role of technological action is then to solve the problem if a platform becomes dysfunctional. Technology is the conscious reflection on habitual techniques, the creation of new ones, and experimentation with solutions—inquiry into technical activities. Although Hickman denies explicit normativity in his typology, it is evident that according to him there hardly can be excessive employment of the fourth category. Most problems of technoscientific modernity result from leaving the technical action without the shepherding eye of techno-*logy*, argues Hickman.[30]

The aim of Hickman's fourfold classification of human activities is to naturalize technology. "Naturalization" here refers to understanding technology in the context of evolution. Dewey's evolutionary concept is "adjustment." Naturalization of technology thus means interpreting it as the human form of adjustment to the environment. Adjustment can take place in two different ways: either the environment is actively worked on to alter it or a being "passively" accommodates its ways in the given situation.[31] Concerning accommodation, Hickman refers to "technologies of the self" as treated by Michel Foucault, Max Scheler, some schools of Buddhism, and Dewey. Dewey describes such accommodative work in the following manner: "It is a change *of* will conceived as the organic plenitude of our being, rather than any special change *in* will."[32] Hickman argues that his four kinds of activities need to be seen as phases in human adjustment. If no adjustment is needed, the non-reflective action is appropriate, but reflection—techno-logy—becomes requisite when existing patterns of action cease to be functional.[33]

Naturalization of technology arises from the general pragmatist philosophical stances. Naturalized understanding of technology, moreover, works well in a practical moral, political, and ecological sense. It involves an ecological continuity among living beings—including humans. This appears, at face value, as a desirable premise given that many of the problems resulting from technology are related to the imbalance of ecosystems that has resulted from humans setting themselves apart from the rest of nature. Furthermore, pointing out that human evolutionary adjustment consists of both active, outward modification of conditions and "passive," inner work on habits and willing relieves Deweyan pragmatism of the charges of favoring one-sidedly dominating action upon nature.[34]

As I noted above, inquiry is nearly synonymous with Hickman's 'technology.' Citing Dewey, he defines inquiry as follows:

> *Inquiry is the controlled or directed transformation of an indeterminate situation into one that is so determinate in its constituent distinctions and relations as to convert the elements of the original situation into a unified whole.*[35]

Thus, inquiry is the human conscious effort to control and manipulate the conditions of life. Since inquiry is always oriented by the goal of transformation, i.e., somehow a *better* situation that can be worked out from the present conditions, it is essentially technological, never purely contemplative or observational knowledge acquisition. At the same time, however, inquiry is a conscious and reflective pursuit: it does not impose blind action on reality. In this manner the notion of inquiry has both technological (*poietic*) and scientific (theoretical) facets. In these descriptions Dewey and Hickman obviously echo Peirce's and James's earlier accounts of inquiry.

Hickman further clarifies Deweyan inquiry through five considerations. He first reminds us that Deweyan inquiry rejects scientific realism. Theories and concepts are tools in solving the problems of life; they do not primarily reflect "a pre-existing reality." In other words, facts are always case dependent, and cases alter as new problems arise—no fixed set of scientific problems exists. Second, inquiry refers to the strategic, general method of intelligence. Inquiry employs a vast array of tactical means to achieve its goals. These tactical methods are not fixed but are open to revision. Third, they in fact can evolve and devolve in the course of history. Modern technoscience has meant forceful evolution, while the period between the collapse of the Roman Empire and the early Middle Ages was a time of devolution. Fourth, Hickman reminds us that there are broad areas of life within which inquiry works only a little or not at all. Phenomena from these areas, however, may provide material for inquiry. As examples he mentions dreams, folktales and remedies, insights of mystics, and even crimes. Finally, he stresses that the emphasis in the inquiry is not on the origin of the material but on its analysis and restructuring.[36]

Philosophy forms a special type of inquiry. Concerning philosophy in general, Hickman views its role in culture as "a liaison officer" or translator between different arts and sciences as well as other technoscientific pursuits. Philosophy, however, is no mere cultural facilitator: it has its genuine subject matter and tools, by which Hickman refers to the traditional philosophical questions and methods.[37] Hickman stresses that philosophy has a genuine role among the forms of inquiry, since it addresses ontological commitments in a manner that no other branch of intellectual pursuit does. Yet Hickman is dissatisfied with 20th-century philosophy because of its lack of public role. Following Dewey, he claims that philosophy should address "the problems of men," not only academic issues. The philosophy of technology, according to Hickman, in its subject, is up to this task. He, however, stresses that regional approaches to technology, such as bioethics or the philosophy of artificial intelligence, do not suffice; a "philosophy of the technological culture" is also needed. Drawing on Peirce and Dewey, Hickman demands that philosophy have a public role in taking the culture and people beyond mere immediacy—unreflected conscience in ethics, unstudied taste in arts and media, straightforward religious thinking and emotional responses, and unthoughtful patriotism, for instance.[38] Rethinking conceptual opposites that rely on the foundationalist epistemology forms Hickman's central method in philosophy.[39]

Hickman employs the word "technoscience" to "refer to disciplines that include the natural sciences, various types of engineering, agriculture, and so on."[40] Thus, technoscience in Hickman's vocabulary more or less equals the historical-social phenomenon that elsewhere in this study has been understood as technology or more specifically as modern technology:

> The term 'technoscience' signals the fact that the 'scientific revolution' of the seventeenth century initiated the systematic use of instrumentation for experimental purposes, thus forever conflating putatively 'pure' theory and putatively 'applied' instrumental practice. In the seventeenth century, science transcended its role as 'knowledge' and became forever 'technoscience' or instrumentally experimental knowledge.[41]

Hickman's technology, therefore, contains sciences as its subcategories. Consequently, as we saw, he rejects all forms of scientific realism and claims that even scientific laws are only tools in the process of human adjustment.[42]

In sum, Hickman argues that everything that is perceived as a problem resultant of technological—normally understood—advance is, in fact, a problem related to the empirical phenomenon of "technoscience." Thus, in

Hickman's usage, "technoscience" is the socially descriptive term and "technology" is the existential, logical, and normative term. If some problems exist, the adequate method for confronting them is the renewed, improved inquiry, that is: technological action. The state of affairs in which some problems are grave and some persistent is due to failures in applying the method of inquiry. Failures could, for example, result from false metaphysics or unquestioned interests. The cure for the malaises in technoscience is more and better technology.[43]

Recasting

The central strategy in Hickman's approach is to recast traditional conceptual distinctions.[44] Following the classical pragmatist lead, Hickman observes that oppositions that often remain unchecked generally lead philosophy into *cul-de-sacs*. More particularly, unexamined notions underlie the fact that philosophy has been ill able to analyze technology and propose constructive solutions for the problems humankind encounters in the technoscientific phase of its history. In this section I will summarize Hickman's recasting of conceptual distinctions. Some of these questions will be revisited later in this chapter.

As I have already pointed out, the starting point for the pragmatist reevaluation of the philosophical tradition is the critique of representationalist epistemology. The crux of the argument is that since human life constantly moves forward and sets goals for itself, "the spectator theory of knowledge" that centers around a static subject and its picture-like relationship to the world is proven inadequate. In the context of the philosophy of technology Hickman develops this theme as the criticism of the practice-theory distinction. He argues that theory and praxis are always interwoven. Knowing must be understood within its practical context: concepts and theories develop in the process of inquiry that gets its stimuli from real problems of living. Inversely, practical endeavors most often have an ingredient of reflection in them. Hickman perceives theories as tools and acknowledges that tangible tools have intellectual meanings built into them.[45]

If knowledge is consistently understood in the context of praxis, it comes to be seen in a thoroughly relational manner. The search for absolute certainty appears as an illusory idea, since all solutions are more or less temporary and open to revision—as they depend on conditions that may change. Nor is there any fixed point from which to have a totally unbiased, objective picture of the world. This understanding should lead one to humility in light of fallibilism, not taking any research result as final or exhaustive, and to active, incessant learning. Concerning the grounding of knowledge, accepting the relationality means anti-foundationalism: no firm foundation for knowing—be it the presence of self-consciousness, *a priori* ideas, or pure sensuous impressions—exists.[46]

Many of the pragmatist moves of recasting exhibit an anti-dualistic inclination. Dualisms such as phenomenon-noumenon, body-mind, eternal-temporal, transcendent-immanent, duty-inclination, fact-value, consciousness-external world are not left unchecked, and the tendency is to claim that alleged opposition is artificial.[47] Pragmatists generally aim at interpreting the distinctions rather as functional than as substantial or ontological. Leaving the deepest metaphysical questions aside, in the context of the philosophy of technology anti-dualism entails recasting the distinction of mental and material. Thus, for Hickman, in addition to devices and technological systems, concepts, theories, and procedures of science, education, and democracy, as well as public institutions, are technological tools.[48]

The pragmatist project of naturalization forms a strong anti-essentialist line of thought. As theories are tools they do not mirror any permanent essences: numbers or natural laws, for example.[49] Nor does a prefigured human essence or nature exist; what human beings in reality are is a matter of inquiry. The same applies to technology: no fixed essence characterizes it, but tactical methods of technology are diverse and open for revision.[50] Hickman takes this kind of empiricism and critique of realism far and emphasizes that Deweyan "radical empiricism" grants an important status for the non-cognitive experience: "it offers access to relations within the focal areas of experience as well as to vague areas at its fringes."[51]

Notions of nature receive pragmatist treatment as well. Hickman makes a distinction between "nature-as-nature" and "nature-as-culture." On the one hand, he characterizes the first as nature encountered without mediation—instrumentation or reflection—and the second as 'nature' resulting from conscious work on values and meanings related to nature.[52] On the other hand, elsewhere he emphasizes that nature is always a social, or cultural, construction:

> In fact, productive pragmatism doesn't treat nature as a 'thing' at all, but as human social artifact constructed from many different received ideas, interests, test results, points of view, and working hypotheses. Dewey thought that one of the most serious difficulties associated with Aristotle's view of nature was that he had treated natural science as an empirical activity rather than as an experimental one.[53]

Consequently, some unclarity or tension appears to exist in Hickman's rethinking of nature. The point concerning radical empiricism would, however, suggest that some non-cognitive, immediate access into experiencing nature was available. In any case, Hickman emphasizes that the distinction between nature-as-nature and nature-as-culture is not ontological but functional; there are not two separate realms, only two ways for humans to encounter nature.

If 'nature' is a human construction in Hickman's understanding, public realm is even more so. "There is no such thing as a 'natural' public," he claims.[54] Beyond mere triviality "publics as products" are understood as follows. Quite concretely, public problems in a modern world are embedded in technology and material culture, although abstract philosophy and political theory often succeed in being unaware of this embeddedness. In an obvious way this is manifest in communication technologies that connect individuals and create public spaces. Telegram, telephone, the Internet, and cellular phone create unique conditions of being communal—different platforms for the possible publics.[55] As an aside, Jonathan Glover in his *Humanity* (1999), while analyzing the First World War and the Cuban Missile Crisis, pays attention to how the technologies of communication both enabled and set the limits for negotiations on the brink of a global conflict. An awareness of technological mediation was an element in solving the Missile Crisis. The Soviets, for instance, used a TV broadcast to announce their conciliatory position early enough. The diplomatic steps that led to the First World War were, to an extent, marked by the slowness of communication technologies, unawareness of modern technology's character, and "technics-out-of control"—both in reality and as an argument for the inevitability of the war.[56]

As the previous chapter suggested, technological embeddedness concerns other and more subtle phenomena than communication alone. Hickman states the point: nation-states arose not out of sophisticated and well-honed political theories, but out of the need to control and regulate the new forms of technological production.[57]

Hickman notes that the liberalist tradition from Locke to John Rawls and Robert Nozick (1938–2002) has ignored the role of technology in shaping public life.[58] Further, he argues that functional concepts private/public work better in solving societal problems than structural, ontological concepts individual/society. The functional viewpoint concentrates on what is to be done in the present as we face the future, not on overall analyses of the society as a whole. Following Dewey's lead, Hickman stresses that the individual, in fact, is an abstraction, since in reality there are no individuals before they are conceptually separated from their social contexts. Therefore, he finds it erroneous to use conceptual tools such as individual and society without acknowledging that they are tools, not existing entities. This criticism has consequences concerning the ideas of a social contract made by individuals.[59]

> Dewey thought the concept of individuality advanced by the empiricists, by those of the 'classic' school no less than those who are our contemporaries, an empty abstraction. He argued that individuals are identifiable as such only

> after being abstracted from historical and cultural contexts. They are neither found, nor could they find themselves, in a state of nature prior to associations with other human individuals. Nor are they the bearers of absolute natural rights apart from and independent of any human society. The reason for this is a technological one. There are no individuals before there is communication, for communication is a technological artifact that is a necessary condition of awareness of oneself as an individual.[60]

The last of the recasting moves that still needs attention concerns means and ends. Hickman argues that means and ends can be separated only abstractly: "In the production of every successful artifact, which is to say in every stage of successful inquiry, means and ends so interpenetrate that they can be sorted out only in retrospect."[61]

Hickman emphasizes that means and ends "receive checks from one other" and that both means and ends in themselves must be open for revisionary inquiry. "Neither ends nor means are privileged, and conclusions reached are always susceptible to further inquiry in the event that new doubts should arise, as does, in fact, frequently occur."[62] In other words, Hickman denies the existence of any transcendent, absolute, or fixed ends—the classical search for the *summum bonum.* Fallibilism reaches to ends as well; they are merely "ends-in-view."[63] To elaborate on the dangers of fixed ends Hickman reviews cases in which inquiry—in its forms of both art and science—has been subjugated to ideological ends. Soviet science and forms of religious fundamentalism are his favorite examples.[64] Hickman, moreover, points out that the view that means are somehow inferior to ends is based on prejudices: in pragmatist understanding, ends are not superior to means. In real life, it follows that "fine arts" are not separable and no better than "practical arts."[65] In understanding technology, pragmatist recasting of the means-ends distinction reveals the untenability of "straight-line instrumentalism," the view that technological solutions are neutral means for freely chosen ends. Since the strict distinction of means and ends cannot be made and since "the means" considerably shape human life, as in the case of the publics, no grounds exist for talk about the neutral means.[66]

Pragmatism and Other Philosophical Traditions

Based on the aforementioned pragmatic moves of recasting, Hickman puts forth a critique of present philosophical traditions, especially in their attempts to address technology. Of analytic philosophy, Hickman notes it has in general failed to grapple with technology as a fundamental cultural force. He claims that analytic philosophy has been too abstract to do this and highlights three features of this emphasis on the intangible. First, analytic philosophers have focused so much on language, conceptual and "ordinary," that other kinds of tools, instruments, and material objects have remained out of their sight. Second, Hickman argues that despite Wittgensteinian attempts at undercutting the assumptions of substance-accident metaphysics the present analytic philosophy firmly stays within its sway. Hickman's third claim is that analytic philosophy has by and large neglected the human body and the embodied and situated character of perception and knowing. He finds it ironical that once frenetically anti-metaphysical tradition currently appears to be the most metaphysical in contemporary thought.[67]

Concerning other major traditions Hickman does not launch such a comprehensive critique but makes more detailed comments on their analyses of technology. I will highlight a few of these comments here. Although Hickman acknowledges the merits of Heidegger and Maurice Merleau-Ponty (1908–1961) in examining the roles of tools in concrete human life, he notes that their contributions to ethics, politics, and the reform of technological culture are either weak, absent, or misguided. He perceives Heidegger's favor for National Socialism and Merleu-Ponty's Marxism as wrong tracks and their less political phases either as romanticism or weak liberalism, respectively.[68] The first-generation critical theorists Max Horkheimer (1895–1973) and Herbert Marcuse (1898–1979) Hickman criticizes for pessimism ensuing from their wholesale critique of technoscience as an ideology—an approach that ill enabled them to propose concrete reforms. Of Jürgen Habermas's second-generation version of critical theory, Hickman notes that it retains the traditional distinction between technosciences (that deal with "facts") and humanities (that address meaning), which is not fruitful for an adequate treatment of technology. Moreover, Habermas has not addressed the questions of technoscience forcefully since the 1970s. Hickman counts Langdon Winner in those drawing on the legacy of critical theory. Winner's argument that in contemporary society technological institutions are more and more replacing or molding the political ones receives Hickman's approval. Winner's substantial plea is that this development should be reversed, so that technology would serve politics. For achieving this Winner has proposed, for instance, small-scale participatory democracy. Hickman, however, doubts these methods and calls them "too Jeffersonian."[69]

Andrew Feenberg, a student of Marcuse's, is a contemporary critical theorist, who has created one of the most influential positions in the current philosophy of technology. In addition to appreciating Feenberg's position on its own terms, Hickman claims that Feenberg has moved close to the Deweyan lines of thought.[70] He summarizes the pragmatist features in Feenberg's thinking as follows:

> He has (a) moved from an essentialist to a functionalist understanding of technology; (b) developed a moderate form of social constructivism; (c) rejected a Heideggerian-type romanticism in favor of a naturalized technology; (d) rejected the critical theorists' notion of technology as ideology; (e) ac-

> cepted the idea that the project of Enlightenment rationality is not as much of a threat as the critical theorists had imagined; (f) proposed the idea that technical decisions are made within a network of competing factors in which one weighs various desired ends against one another; (g) warned against the reification of the results of inquiry as if they had existed prior to inquiry (Dewey's 'philosopher's fallacy'); and (h) recast technology in a way that crosses the line between artifacts and social relations.[71]

These shortcomings of the other approaches and his perception of the growing employment of pragmatist thought among the philosophers of technology, Hickman takes as arguments for the relevance, and indeed the superiority, of the pragmatist position in the philosophy of technology.

Technology and Change

Given his anti-dualistic and anti-essentialist starting points, it is fitting that Hickman claims to offer an insightful solution to the often-visited issue of technological change. Although both voluntaristic instrumentalism and technological determinism[72] are undesirable conclusions in understanding the relationship of technology and social change, within the conventional assumptions certain inherent tendencies exist, from which either instrumentalism or determinism easily follow. Hickman addresses the question both in a negative way, criticizing instrumentalist as well as determinist positions, and in a positive way, presenting a Deweyan account of freedom and social change.

Hickman mostly agrees with Langdon Winner's critique of "straight-line" instrumentalism—the view that technology in its different forms is a neutral means for ends that are independent of it and freely chosen by humans.[73] This critique has the following features. First, Winner points out that the assumption that end valuations for technological actions would come from a non-technological realm is erroneous. At least in modern technological settings such a separation of technological means and non-technological ends becomes impossible. This is due to both the manifest and the concealed complexity of technological systems, which makes their instrumental control very difficult.[74] In addition to Winner's empirical point, Hickman has his own conceptual critique of means-ends distinction; he points out that straight-line instrumentalism relies on some metaphysical or scientific realism.[75] Second, Winner points out that a technological system is not a mere combination of technical means; it is a real, active force in itself. "The system possesses inertia and goals of its own," as Hickman paraphrases Winner.[76] Third, this autonomous character of technological systems may lead to "reverse adaptation." This phenomenon can be summarized as follows:

> Technological systems become severed from the ends originally set for them and, in effect, reprogram themselves and their environments to suit the special conditions of their own operation. The artificial slave gradually subverts the rule of its master.[77]
>
> *[R]everse adaptation*—the adjustment of human ends to match the character of available means.[78]
>
> Efficiency, speed, precise measurement, rationality, productivity, and technical improvement become ends in themselves applied obsessively to areas of life in which they would previously have been rejected as inappropriate.[79]
>
> A person working at a job that requires high standards of exactitude may find that those standards begin to encroach on other areas of his or her life.[80]

Fourth, related to reverse adaptation, Hickman notes that a common conception of freedom that finds it in the unhindered expression of impulses is inadequate, since the conception ignores the fact that impulses themselves are to an extent products of social environments.[81]

Following Winner in his criticism of the instrumental understanding of technological change, Hickman drifts toward the ideas of technological imperative, autonomous technology, and technological determinism—ideas that are as foreign to him as instrumentalism.[82] Although the notions are distinct, Hickman focuses his attention on technological determinism, which summarizes the allegation that human freedom in technological change is limited, or indeed absent. The idea of technological determinism has many variations. The core of the notion, however, is a contention that technological artifacts, knowledge, and systems have complete laws of development, which give little room for human freedom, and that technological development greatly steers other social and cultural development.[83] Hickman raises Marxism and Ellul's critique of technology as the primary examples of technological determinism.[84]

Hickman's Deweyan arguments against technological determinism in general are as follows. To begin, he notes that the viewpoint from which technological determinism is usually conceived is an abstract panorama. He, by contrast, favors taking concrete and involved human beings as the starting point.[85] Second, Hickman observes that Marx and Ellul are naïvely realistic about the notions of cause, effect, necessity, and freedom. In pragmatist accounts such concepts are analyzed within the context of inquiry—they are tools of inquiry, not substantial elements of the world that preexists inquiry. According to Hickman, this reading applies to all levels of reality. In the case of technological determinism it means that no fixed "laws of history" exist. Instead of being mere "past-looking" excavations of necessities, historical-social interpretation is a part of "forward-looking" life as well.[86] Third, he points out that both in Marx and Ellul a tension exists between the deterministic understanding of the historical process and the pleas for the conscious groups to steer the societal development. Visions of a desirable future cannot be meaningful if full-blown technological determinism holds sway.[87] Fourth, Hickman criticizes Ellul in particular for reifying the notions of *La Technique* and technological system that should be mere tools of his analysis of the modern technological condition. Moreover, he observes an ambiguity in Ellul concerning where the power lies: sometimes it appears to lie in a technocratic establishment, sometimes the system itself appears to execute the power.[88]

While addressing the issue of technological determinism Hickman visits Dewey's account of human freedom in general. His key point in favor of human freedom in action arises from the human awareness of temporality and the use of symbols. Drawing on Dewey, Hickman argues that the symbolic thinking that takes place in time raises human beings above mechanistic processes. The choice, however, is not fully a human privilege in Deweyan naturalistic account: there is selective behavior in all nature, beginning with atoms. The basis of human freedom, therefore, lies in this preferential choice common to all nature. Humans, however, possess the unique ability of forward looking via symbolically conceived time, especially the future.[89]

Hickman, nonetheless, takes the "deterministic" critiques of technological society somewhat seriously. He admits that situations and political conditions exist in which the human power to act according to free choice is disabled by technoscientific production and products. He, however, refuses to take this as an inherent feature of modern societies and the technologies they employ. In Hickman's interpretation, at the heart of an unfree condition always lies an interrupted inquiry. He agrees with Winner that technology gaining the upper hand over politics is an undesirable situation. In Hickman's translation of Winner's terms, "technology," however, is substituted by habitual techniques, and "politics" is understood as technological work on the publics.[90]

Although Hickman argues for the possibility of certain freedom in technological change, he criticizes the underlying assumptions of the instrumentalist views. Hickman argues that the mainstream idea of science as "value-free" theoretical pursuit detached from practice as an underlying assumption obstructs the adequate understanding of technoscience as a human pursuit. Hickman compliments other praxis philosophies, phenomenology and Marxism for instance, for paying attention to the practical situatedness of human existence. Yet he observes that these philosophies usually just reverse the order of practice and theory: practice is fundamental and primary and theory only a modified form of primordially practical being in the world. Pragmatism, in Hickman's reading, nevertheless, recasts the distinction altogether. He, in fact, claims that the pragmatist understanding perceives the theoretical, practical, and productive aspects in all human action. Even the most theoretical activities produce something: meanings and new conceptual tools, for example. Moreover, as we noticed previously, Hickman rejects the distinction between morality and knowledge. Inquiry is the morally laden pursuit of goals within the forward-moving human existence, and morality has no privileged sphere outside empirical research. In other words, *epistēmē*, *phronēsis*, and *technē* are seen as interwoven.[91] Further, since all human pursuits are inherently *poietic*, productive, Hickman calls the Deweyan approach "productive pragmatism."[92] In Deweyan pragmatism science and technology are understood together as forms of inquiry that make use of technological artifacts, both abstract and concrete; technological change is as well understood as a social process that consists of agents, existing artifacts, methods of inquiry, and interacting preferences. Thus is created a position that Hickman claims to be inherently free from the traditional problems in understanding technological change.[93]

PRAGMATISM AND THE ETHICS OF TECHNOLOGY

I have so far elaborated on Hickman's philosophy of technology and the central pragmatist reformative moves within the philosophical tradition in general. Now it is time to evaluate the consequences of these stances in the ethics of technology. In this section I will first look at what Hickman perceives as the current ethical-political malaises and their causes. Then I will summarize his position concerning facts and values, after which I will proceed to describe his understanding of the ethical inquiry.

Malaises in the Technoscientific Society

Hickman frames the problems that he perceives as malaises in the technoscientific society within his overall Deweyan philosophy. These malaises exist within three main spheres of understanding: science, religion, and citizenship. First of all, Hickman argues that there are malaises in technoscientific practice and in ways of understanding what technoscience is. He characterizes a view of "Enlightenment"[94] technoscience that he claims still to be prevalent. This view wants to ground science in indubitable foundations, believes in inevitable and linear progress, and perceives nature as an adversary to be dominated and mastered. Against this view, which Hickman perceives as outdated, he sets Dewey's "post-Enlightenment" thought: fallibilism, radical empiricism, relationalism, and evolutionary naturalism as the correct ways of understanding science. According to Hickman, the Enlightenment view is not truly open for inquiry, for instance, due to its reliance on foundationalism. It is, moreover, more prone to lead into ecological crises than the pragmatic understanding.[95] Hickman, however, appreciates technoscience and its boons. He therefore criticizes postmodern thinkers who blame technoscience for having gone astray and relativize its epistemic and moral worth. Hickman notes that Richard Rorty's "postmodern" reading of Dewey is tendentious.[96]

In addition to the general understanding of technoscience, flaws in its practice form another set of issues. Scientific fraud, questionable practices such as patenting life-forms, policies that have led into decline in public health, and the cost overruns in big publicly funded projects such as Texas Superconducting Super Collider are Hickman's examples. In all these and others, the flaws are not in technoscience itself but are due "to the failure to apply its finest and most appropriate methods in a consistent manner," argues Hickman.[97] This can result from misunderstanding the character of technoscience; subordinating the public good to a professional, class, or private interest; technoscience as body of knowledge toppling technoscience as method of inquiry; or the reliance on tradition instead of seeking reforms.[98]

When addressing religion, a major reservoir of tradition, Hickman spends time in analyzing fundamentalisms,[99] Islamic and Christian, as phenomena that powerfully obstruct the employment of inquiry. Although fundamentalist doctrines are cruder than the philosophical foundationalism, absolutism, and transcendentalism of which pragmatism is critical,[100] they form an extremely potent social force. Hickman notes with the help of statistics the presence of the fundamentalist thinking at schools in the United States. The themes that he finds especially troubling are "the creation science"[101] and sex education. Hickman points out that in many respects Christian fundamentalism, under the rubric of religious belief, cherishes scientific and intellectual elements from the early modern period. Drawing on Northrop Frye's work he argues—correctly in my knowledge—that literal-demotic reading of the Bible is no archaic Christian practice, from the first century C.E. for instance, but derives from the era of modern science. According to Hickman, the literal reading does violence to the mythic, metaphoric character of the *kērygma* ("proclamation," Jesus' and his apostles' teaching) and is of course intellectually inconsistent, if it is believed to be "the original" interpretation.[102] Hickman makes a similar remark concerning creationism: the creationists desire to conserve biological thinking that became inadequate 150 years ago. The oxymoronic name "creation science," in fact, points out that one needs to make use of some science to form such a combination. Hickman's point, in other words, is that both in the interpretation of the Scriptures and in creation science Christian fundamentalism relies on static and old-fashioned science—not on genuine spiritual insight. Further, Hickman describes how the fundamentalist moves in sex education include scientifically inadequate information concerning safe sex, abortion, and homosexuality. Of course, an even greater flaw than these is that fundamentalism, as a general view of life, actively and explicitly wants to suppress inquiry, at least in some areas. And when this takes place in education it is harmful concerning not only the body of knowledge but especially learning the method of inquiry, so goes Hickman's argument.[103]

Education and citizenship form the third central sphere in which Hickman observes malaises: "one of the more interesting paradoxes of our technological milieu is the conspicuous fact that the defining methods and outlooks of the technosciences, despite their remarkable successes and widespread use, have been a subject of popular misunderstanding and even disparagement."[104] Counting out religious ideologies and economic interests, Hickman observes a lack of correct information about technoscientific projects at the root of this problem. He reviews studies on the public knowledge on the thinning of the ozone layer and acid rain as examples. In these studies the great majority of the U.S. population could not identify the causes of the two phenomena. Culprits for this lack of information—and misinformation that exists as well—are the fragmentation of scientific disciplines, bad education, and the hesitance of professional people to communicate their research results to the public and exhibit their wider social implications. This last point is related to the Enlightenment understanding of science and technology as separate, perceiving science as producing individual "facts" without any direct practical meaning or value.[105] Hickman takes the fact-value distinction as one more target of recasting.

Facts and Values

Hickman is critical of ethical approaches that attempt to ground ethics on any non-relational basis; he does not accept any transcendent goods or standards of right. Furthermore, in the domain of ethics, Deweyan pragmatism means reconsidering the fact-value distinction that has been central in modern philosophy. If there is no theory that is free from practical contexts, there cannot be pure facts, and hence the value component is interwoven in all factual descriptions. In other words, all knowledge has a value dimension, a moral dimension.[106]

To elaborate on the value-ladenness of knowledge, Hickman presents the following considerations. First of all, facts are not given but selected. If one, for instance, notices that something is worth paying attention to or noteworthy, it is a judgment with a value ingredient. Therefore, the choice and framing of facts rely on valuations. Conceptually, facts, in pragmatist understanding, are always facts of a case; in other words, a problem and its solution form the structure of knowledge.[107] Hickman, furthermore, stresses that human perception is not passive absorption but active exploration:

> The image of vision that Dewey presents is not that of a ray entering the eye from an object, but rather that of a probing tool that moves and checks and interacts.[108]

Second, concerning discursive forms of knowledge as distinguished from mere acquaintance, Hickman argues that knowledge, in fact, results from production, which obviously means that it is a fruit of a process oriented by some values.[109] The inverse holds true as well: "Technoscience is a moral enterprise because it is capable of influencing the goals and ideals by which people live."[110]

Third, Hickman makes a historical point concerning modern science. He notes that technical instruments central to modern science, manipulative experimentation, and functionalist ontology—as opposed to substance-accident metaphysics—increase the productive character of modern knowledge acquisition, compared to its historical predecessor, the Aristotelian contemplative and observational science.[111]

Looking at values, Hickman emphasizes that fallibilism applies to them as it does to everything else. Since no transcendent source of values is taken into account, it follows that values are the results of testing. Therefore, they need to be revised and clarified through inquiry that is both empirical and reflective.[112]

> Values arise out of inquiry, and once they are refined by inquiry they are brought back to the situations from which they originated in order to ascertain whether they are appropriate.[113]

Ethical Inquiry

Although Hickman explicates it little, ethics can be fruitfully placed within his schema of four kinds of activities and his broad understanding of technology. Then the loci of ethics are the second (cognitive, non-instrumental) and fourth (cognitive, instrumental—*technological*) categories. Of course, all human action is morally bound. Ethical reflection, however, can take place only in cognitive modes of action. Since the second category is narrow, ethical *reflection* is practiced mainly in the fourth, technological category. Moreover, since most of human life happens in the habitual, non-reflective categories (the first and the third), moral *life* occurs in them. Consequently, pragmatist ethics emphasizes ethical reflection on habits.[114]

> Taking his cue from William James, Dewey argued that habit was both mainspring and flywheel, both inner tension and live inertial weight, of human action. *Habits* bind us to old ways of doing things and make us reluctant to initiate novel courses of action. But they, and especially the *shared habits we call 'institutions,'* also provide the springboard from which new discoveries can be made and new paths of action can be entered upon.
>
> Habit and thinking are thus for Dewey not polar opposites, but phases within human experience. Habit determines the channels within which thought operates, and *thinking takes place within the fabric of habits.*[115]

We mostly act habitually. We have habits of feeling pleasure (the first category) and using tools (the third category). When we act habitually, the act is more or less guided by a habitual pattern. Consequently, it is not adequate to attribute too much moral freedom in habitual acts. In pragmatist ethics the freedom appears to exist mostly in the reflection on habits. This is the pragmatist recognition of the prevailing non-cognitive parts of life and the strategy for placing cognitive morality in the fabric of life.[116]

In the pragmatist account ethical concepts and theories do not differ from those of science and engineering. Ethics is one form of technological inquiry. Ethical concepts are tools as well as any other concepts:

> Dewey's example is the Golden Rule. Like any other tool, it gives no knowledge in and of itself concerning what should be done with it. But as a tool of analysis, it is like a finely constructed axe: it 'helps me hew and straight and fine in clearing out this jungle of relations of practice' (EW3:101).
>
> Modifying his metaphor somewhat, Dewey claims that moral rules (hypotheses) and scientific hypotheses are like another kind of tool: a nautical almanac. They are the aids of navigation, but they do not tell us where to sail.[117]

The method of inquiry is to be applied both to the solution of moral problems and to the assessment of ethical tools. Both are to be solved experimentally. Thus there are no values or norms that stand outside the realm where critical inquiry is practiced.

> Moral action does not need to be justified by a transcendent ground of values; neither is it unjustifiable. Moral action is, quite simply put, what is most efficient and appropriate to the resolution of a given problematic situation.[118]

If ethical maxims or things that are perceived valuable do not pass the tests any longer, they must be replaced by new ones that work better. Testing and experimentation is, however, understood broadly and flexibly as experiments in "the laboratory of life."[119] An example that Hickman has treated at some length is the same-sex marriage. Hickman argues that research data prove the traditional taboos ungrounded and suggest that morals concerning marriage must be rethought.[120]

Hickman claims that moral theories and theories of physical science are similar in form; both are hypothetical: if x, then y. He, furthermore, professes a clear view that ethical theory is indispensable:

> But there is also a mutilation involved in claiming that moral activities are possible in the absence of moral hypotheses or theories. The alternative to hypotheses as a basis for action is impulses. Actions based on impulse may serve; but if they are able to do so, it is only a matter of luck.[121]

In sum, Hickman's thinking on facts, values, and ethics is intellectualist, consequentialist, and naturalist. For him, ethics is just one form of inquiry. Ethical concepts and theories are tools, the functionality of which needs to be tested again and again as the inquiry progresses. The form of ethical notions is hypothetical: "if this, then that."

RESPONSIBLE TECHNOLOGY

Hickman's analysis of technology, his perceptions of the malaises in late modern society, and naturalist ethics come together in his reform proposals, which he summarizes with the term "responsible technology." I will first visit Hickman's treatment of grand historical scenarios and then proceed to his concrete proposals.

Technology and Grand Narratives

In addition to rejecting instrumentalist and determinist ways to understand technological action and change, Hickman discusses myths in framing humankind and technology in a temporal continuum. Following an ancient cue he acknowledges that, to an extent, mythic understanding is inevitable in linking the ruptures of *logos* and in envisioning large historical and existential narratives, the particulars of which cannot be fully known. Hickman characterizes Jacques Ellul's and Martin Heidegger's views on technology as this kind of myth.[122] Simplifying crudely, the common point in both Ellul's and Heidegger's narratives is the idea that humanity has become trapped in its project of acquiring control over reality. Technology is not devices, tools, or means to be chosen but an underlying condition of being, knowing, and acting. Therefore, no way out of technology appears to exist. Attempts to control technology take place within the same technological framework. Ellul observes religion as a part of this development: technology has acquired the character of an idol and institutionalized religion resembles technology in its ways of bargaining with God. Heidegger's answer for the human being in this condition is waiting, detachment.[123] In the posthumously published *Der Spiegel* interview Heidegger stated that technology will destroy humankind unless a god saves it. This saving could refer, for example, to a new disclosure of Being or an event in which an absent deity becomes present. Humans can do nothing to promote this possible salvation, since it must be something that comes from outside of the present ways of encountering reality.[124] Ellul's views were articulately Christian and somewhat more active than Heidegger's. Ellul described the correct response as "hope." Hope is the opposite of the expectation central both in technology and in superficial religiosity. Expectation refers to anticipation of a certain empirical, i.e., intra-worldly, state of affairs. Hope, to the contrary, does not look for occurrences within the world but consists of confidence and trust independent of the worldly affairs—"transcendental" trust. Since technology reaches for the total control of empirical conditions, hope is "the only standpoint outside the technological system." In relation to technology, the Christians living in hope should critically engage in technology in order to strip the idol's mask. In relation to God, hope is an absurd reaction: "man's answer to God's silence." Ellul's hero of hope is Job.[125]

Hickman argues that Dewey's narrative of humankind's destiny is more benign than quietism and religion. Dewey refrains from wholesale judgments concerning technological modernity. His approach emphasizes piecemeal reflective work:

> Dewey's account is quite different. Instead of abandoning a myth of control for a religious myth that turns out to conceal yet another myth of control, his account is a myth of control in which loss of control is not massive, not due to the hegemony of out-of-control technology. For Dewey, loss of control and securing it again is a part of the rhythm of organic life. For Dewey, technology is not an enemy. It is the means of putting things right that have gone sour or gotten off the track.[126]

While "hope," "salvation," and "waiting" are apt words in the context of Ellul and Heidegger, Dewey's approach is characterized by "responsibility," claims Hickman.[127]

Practical Reform

Returning to the malaises reviewed earlier, Hickman presents Dewey's three-point program of remedies. The points are distancing "mainstream" religion from fundamentalism, scientists' work on social reform, and the proper education and participation of citizens. First, concerning religion, Hickman is sympathetic to Dewey's somewhat radical hopes for "a next evolutionary step" within religion. The step would lead religion in a more experiential and reflective direction. Although such a step is not any more on the horizon now than it was in Dewey's time, Hickman argues that public consequences of religious beliefs should be submitted to public testing.[128]

Second, Hickman demands a socially active role for the professionals: "Technoscience is a moral enterprise because it is capable of influencing the goals and ideals by which people live."[129] The active role of the professionals is interwoven with the third point, the role of citizens in the technological society. According to Hickman, the professionals ought to refrain from making use of their authority or hermetic modes of communication—especially in their own interest. Moreover, they should acknowledge that no science exists separate of technology[130] and their common moral implications. Concretely, these demands refer to "public understanding of science," ways to integrate the public into research activities of the society and enhance "scientific citizenship."[131]

Education is the most important vehicle for advancing these goals. Following Dewey, Hickman laments "a loose, scrappy, and talkative education." On the one hand, mere participation and opinionated extroversion are not sufficient for learning. Deweyan pedagogics, on the other hand, also reject hierarchical educational structures, in which bodies of knowledge often serve as backups for authority and obstruct open inquiry. The middle way is characterized as "the scientific attitude," "truly intellectual curiosity," "love of thinking," and "interest in problems as such." These attitudes and approaches should be the goals in educational institutions. Hickman envisions education as a lifelong project. This project, moreover, is not merely epistemic:

> The reconstruction of technoscience for a life curriculum must involve the rejection of the putative gulf between moral and non-moral knowledge.[132] . . . [T]he teaching of 'values' requires no special place [in the curriculum] since, properly understood, such instruction permeates the entire curriculum and infuses it with meaning.[133]

These moral, educational goals are pursued within political and economic contexts. Real education cannot take place without necessary material conditions. Material scarcity—hunger, illness, and fear—disable learning. Learning, inversely, is necessary for real democracy; Hickman quotes Dewey: "The fundamental defect in the present state of democracy is the assumption that political and economic freedom can be achieved without first freeing the mind."[134] Hickman is trustful that public, experiential, reflective, and democratic processes of inquiry are capable of reforming values and creating new ones. He quotes Dewey:

> A culture which permits science to destroy traditional values but which distrusts its power to create new ones is a culture which is destroying itself.[135]

Education, democratic participation, reform of public institutions, improved research, reevaluation of underlying narratives—all this, in Hickman's conception, is technological action. Therefore, his solution for making technology responsible is not less technology but more and better technology.

CRITIQUE AND REFLECTION

Hickman's achievements are evident. He, on the one hand, manages to show that the Dewey corpus contains a complete philosophy of technology that addressed many central questions decades prior to when they were taken up under the rubric of the philosophy of technology. On the other hand, Hickman's work quite successfully points out how the pragmatist approach through its general critique of the philosophical tradition opens genuine, insightful paths in the philosophy of technology. In this way pragmatism manages to cut through conventional dichotomies and dead ends. In what follows I will look at Hickman's position critically. Although problems and shortcomings now rise to the fore, they need to be seen in balance with the other parts of the chapter and Hickman's achievements mentioned.

The Insubstantiality of Ethics and the Ability of Critique

Pragmatist ethics appears thin and insubstantial. Many have worried about this since the times of Dewey. Max Horkheimer, for instance, criticized Dewey for advocating instrumental reason, lack of criticalness of industrial practices, and emphasizing practice over theory and means over ends.[136] Albert Borgmann has put forth a critique of Hickman's position along similar lines. Borgmann, moreover, thinks that while Hickman has managed to clarify Dewey's terminology in drafting a comprehensive philosophy of technology, in this process he has "flattened" Dewey's thought—especially in ethical-political respects.[137] The crux of his critique goes as follows:

> If ends are as variable and adaptable as means, then the critics whom Hickman cites as complaining that productive pragmatism 'is too weak to provide adequate guidance for difficult decisions' are right. In fact, absent firm values and norms, there is no guidance at all. To put the point differently, if values are testable, then there must be strategic or higher-order values that are firm and tell us whether our tactical or first-order values that are being tested at the moment will serve as effective means for the strategic ends.[138]

To put the problem a third way, Hickman argues that ethical inquiry should aim at increased functionality in improving the moral tools, which would entail functionality being a second-order good, against which first-order goods are measured. This would require more clarification and an argumentative account in support of the role given to functionality. Hickman, however, has rejected the interpretation that would entail a second-order good:

> To this I would reply that pluralism and fallibilism, along with democracy, growth, and a host of other such terms, are more or less vague tools, or ends-in-view. They become determinate only in the course of the type of experimental work that is required when there are hard decisions to be made.[139]

I find Hickman's answers to Borgmann's valid critiques insufficient. In themselves "pluralism," "fallibilism," and "democracy"—almost by definition—offer no criteria for better or worse courses of action and inquiry. Nor do they explain what would motivate and guide Hickman's responsible technology.

So far the critical point has been mostly logical, formal: Hickman does not acknowledge sufficiently the need to elaborate on the final-order values. It is, however, fair to provide a little more content for the criticism. Hence, one may ask: What could be threatened by the pragmatist's fully relational ethics? Answers hardly bring up great novelties. Such, admittedly vague but pregnant, notions as autonomy, personhood, human dignity, or in one word: the other come first into the mind. Examples of technological systems that violate human autonomy and dignity through instrumentalization are not rare.[140] Quite easily one could argue that these refer to something that is transcendent and absolute[141]—something that Hickman wants to remove from ethics. Although in their contingent goals and tones of their writing Dewey and Hickman are humane, no real philosophical obstacle seems to exist for rethinking morality in a manner that would sacrifice the human person in order to achieve higher platforms of functionality. Second, it is unlikely that justice can be fully submitted to a relational method of testing ethical tools in the laboratory of life. Third, an ancient tradition since Plato has argued that Good in itself is a concept of a higher order than other concepts.[142] This, of course, is the line of thought Borgmann joins as well.

Yet one more way of putting this criticism exists, a way that is the most immanent in Hickman's approach. When he speaks of improving the conditions of life, I find it unclear to what extent the situations in life may *in themselves* be problematic—that is: inquiry provoking—and to what extent some *ideal* of betterment is at work in Hickman's pragmatism. What tells us when inquiry and improvement are needed and when "things are all right enough"? What tells us the proper direction of the improvement?

The theme of critical normativity is at stake as well in Hickman and Andrew Feenberg's dialogue, but in a slightly different sense. Feenberg has replied to Hickman's characterization of Feenberg's present position as Deweyan. To begin with, Feenberg compliments Hickman's revealing the role of Dewey as a philosopher of technology. This feature of Dewey had remained largely unknown before Hickman's work. Feenberg, moreover, admits that he has in many points come near to Deweyan thought. He, for instance, agrees on the rejection of essentialism and determinism as well as the demands for more democracy in technology. He, nevertheless, argues that important divides still exist. As this chapter focuses on Hickman I will emphasize Feenberg's critique of Dewey and Hickman rather than Hickman's reading of Feenberg and its accuracy.[143]

Feenberg's disagreement with Hickman's and Dewey's approach, in brief, goes as follows. Feenberg finds Dewey and Hickman to fail to understand "the tragic sense" of the 20th century.[144] Feenberg thinks that although the social constructivist turn has distanced him, and most others, from the dystopian, essentialist, and deterministic critiques of technoscientific modernity, the dystopian perspective still has some relevance in the critique of technology. Dewey and Hickman appear to lack this perspective altogether. Feenberg finds them advocates of an optimistic Enlightenment rationalism and American liberalism that fail to address the existential dimension of the human experience. The dystopian view and the tragic sense characterize the tension between technoscience and the lifeworld, "in which meanings and significance are formed and democratic initiatives nurtured." Concretely this takes place in a technological design that can embody "the demands of a way of life and an image of what it is to be a human being."[145] Put more politically, Feenberg argues that

> technology can be and is configured in such a way as to reproduce the rule of the few over many. Designs which have this effect then also narrow the range of interests and concerns that can be represented by the normal functioning of the technology and institutions which depend on it. This narrowing reinforces an elite power, but more importantly, it causes human suffering and damage to the natural environment. Opening up technology to a wider range of interests and concerns could lead to its redesign in ways that support democratic self-management in the future.[146]

At first sight the difference between Feenberg and Hickman appears to be only slight, one concerning more the tone than the substance. Nonetheless, if Feenberg's critical intentions are compared to Hickman's writing—"There are good grounds for applying the methods that have proven successful in the technoscientific disciplines to social problems"—the difference is more clear.[147] Hickman's technoscience is a closed circuit that swallows ethics and politics within itself, whereas Feenberg still attempts to maintain an existential and political critique of technoscience—a critique that would be independent of the rationality of science and technology.

Problems of Straightforwardness

Hickman's picture of technology within human life appears somewhat straightforward and poor in features. In his depiction life, artifacts, ways of thinking, tools, language, and forms of life made up of these seem surprisingly manageable. Techno-logy is reflective, experimental management of habitual action. It manages habits of action in light of research toward ends-in-view that are more functional than earlier ones and the situations to which they led. This picture overemphasizes control; something is missing. How well can we master cultural meanings with *logos*? To what extent can human life, existence, be brought into the light of consciousness? How well do human spontaneity and liveliness of personal interaction fit Hickman's picture of technological life? Finally, to what extent can forms of life, for example technologies as forms of life in Langdon Winner's sense, be administered by *logos*?

At the risk of being uncharitable, I think that Hickman's voluntarism and belief in control surface, for instance, in this passage that uses Platonism as a misleading straw man:

> what our tools do to and for us is dependent upon choices we make and continue to make, dependent on our relations to our tools, and not upon their 'essence' or 'pure form.'[148]

Borgmann's account of the device paradigm, I think, is sufficient to show that "to make choices" and alter our relations with "tools" is not so simple and straightforward as Hickman appears to suggest.

If no reflective transformation of technologies as forms of life would be possible, then the ethics of technology would obviously be a useless pursuit. So, that is not the point. The point, rather, is that Hickman's method is too comprehensive. His concepts seize too much in order to be able to be concretely transformative. Although in the content of his arguments Hickman values piecemeal reforms and (radical) empiricism, his own methodical framework is inclusive and all-encompassing to the extent that it makes the discussion of concrete piecemeal reform difficult. In comparison, Borgmann's ideas on focal reform are concretely targeted and humble in the sense that they do not attempt to address all aspects of life.

On the conceptual level, even after Hickman's elaborations, his wide definition of technology appears problematic. I find problems in three respects: in its content, in its broadness, and in the fact that it is a general definition. The core of the problem lies in the last aspect. Wittgenstein, in *Philosophical Investigations*, as it is known, criticizes philosophy that aims at such general conceptions. Wittgenstein's central concern is what he finds the erroneous pursuit of a general conception of language. He wants to show that we can study how language works in life and how words have meanings, but we cannot form a general theory of semantics. His approach is very concrete. His well-known particular example examines the concept of game (*Spiel*). We cannot define what a game is without artificiality. Chess, baseball, hide and seek, poker, solitaire, soccer, and push-pin are all games, but it is impossible to capture them in a definition of game. Yet we know what is a game and what is not. Washing dishes, jogging, and reading a philosophy book are not games, while Tetris, Monopoly, and Lotto are.[149]

I think that the same applies to the concept of technique (here used afresh, without leaning on any specific previous ideas). A violinist must use some techniques to play her instrument. A soccer player clearly employs techniques in controlling the ball and in game-making. A novelist makes use of techniques of crafting sentences and of managing the structure of her book. A process of, say, paper production involves a variety of techniques, most of which are acted out by automated machines. Just enjoying candy, taking a walk, or sitting in a tram car involve few or no technical aspects. Now my point is that Hickman's general definition of technology, which views technoscience as a paradigmatic example of technology, makes us see things in a light that neglects the diversity among techniques and generally in human life. It looks *as if* there is a category of technology that consists of problem-centered, cognitive, reflective, and experiential work on habitual ways of acting instrumentally upon reality. But in human life and in our ways to use the words "technique," "technical," and "technology" no such clear-cut category seems to exist. Playing violin is much more than technique: composition, artistic expression, dialogue with tradition, communication with other musicians and audience are at least as important as technical abilities. This is the context of technique in the violinist's practice. Moreover, it may be that cognitive reflective and empirical experimentation (*logos*, -logy) have little role in perfecting the techniques of playing violin. Repetition or imitation, for instance, may be crucial avenues in this case. Finally, perceiving music as production just does not seem apt. Therefore, Hickman's emphasis on habitual vs. cognitive, on instrumentality, and on productivity may not do justice to technique in the context of playing violin—or playing soccer, writing a novel, or many other activities.

Paul B. Thompson, a leading philosopher of agriculture who himself draws on the pragmatist legacy, has, in fact, argued that Hickman's attempts to define technology are not very pragmatic, either. Thompson points out that a pragmatist usually has a provisional understanding of what "technology" means, and if a more precise parsing is needed it should be guided by each particular pragmatic context. Further, as we saw, Borgmann criticized Hickman for "flattening" Dewey's approach.[150] I somewhat agree with Thompson and Borgmann. In the above respects Hickman's approach appears too straightforward, one-dimensional, and totalizing. This is due to Hickman's understanding technology in a broadly human and existential sense but simultaneously setting the technoscientific procedures as the measure of good management in life. Thompson, moreover, has noted that Hickman's position does not come to grips with what is problematic in technology. To an extent this is true: Hickman sanguinely manages not to look in the eye the problems of technology itself; he finds faults in right-wing economic policies, conservative religion, or bad media, but technology in itself is unproblematic.

CONCLUSION

If I express the conclusion of this chapter in a simple manner, it goes as follows: Hickman's approach is rich and insightful on technology, but I am less convinced of its benefits in opening ethical perspectives. For instance, in recasting the persistent distinction between *poiesis*, *praxis*, and *theoria* or in treating the elements of technological change, Hickman's pragmatism appears genuinely to create better ways to implement a philosophy of technology. Moreover, understanding technoscience in a manner in which theoretical, productive, and practical—in real life: science, applied science, development, artifacts, politics, and policies—intertwine appears necessary in most concrete moral problems our civilization faces. Consider, for instance, climate change and other facets of the ecological predicament. Hickman's proposed remedies, improved education, democracy, and research, are not exotic, but by the same token they are the realistically possible means available for liberal democratic societies.

Despite the insightfulness and practical applicability of productive pragmatism, when it comes to ethics, I worry that Hickman's approach lacks orienting and critical power. Philosophy and ethics need not only present solutions that can be put into work immediately. Presenting possibilities, questioning what *is* against what *ought to be*, discussing the demands of the regulative ideals have been parts of the Western ethical tradition. These obviously come into a new light if the pragmatist naturalism and relationalism are accepted: no transcendental or absolute source of value exists to be the benchmark for moral and cultural critique. Nevertheless, without such reference points—reason, justice, person, autonomy, or dignity—ethical argumentation becomes thin and loose. Some substantial commitments are necessary for most practical reforms as well. Consider, for instance, international agreements, environmental legislation, and taxation as tools for steering development. As we saw, Dewey's and Hickman's progressivism appears to lack intellectual backing. This I perceive as the dilemma of pragmatist ethics: if it is truly pragmatist it is ethically thin.

NOTES

1. Hannu Raittila, *Canal Grande* (2001), 119–120, translated by Andrew Chesterman. Raittila's novel depicts a Finnish delegation in Venice. They are part of a larger UNESCO project that aims at saving the palazzos on Canal Grande from the rising seawater. The protagonist who is the narrator of the above passage is a Finnish engineer, Mr. Marrasjärvi.

2. For a programmatic proposal in favor of pragmatism in the philosophy of technology, see Jozef Keulartz, Maartje Schermer, Michiel Korthals, and Tsjalling Swierstra, "Ethics in Technological Culture: A Programmatic Proposal for a Pragmatist Approach," *Science, Technology, and Human Values* (2004).

3. Don Ihde, *Postphenomenology: Essays in the Postmodern Context* (1995). Peter-Paul Verbeek, *What Things Do* (2005), 99–119.

4. H. S. Thayer, "Pragmatism," *The Encyclopedia of Philosophy* (1967).

5. Charles Sanders Peirce, "The Fixation of Belief," *Writings of Charles S. Peirce: A Chronological Edition, Vol. 3: 1872–1878* (1986), 242–257.

6. Peirce, "How to Make Our Ideas Clear," *Writings of Charles S. Peirce: A Chronological Edition, Vol. 3: 1872–1878*, 257–276.

7. Ibid., 266.

8. Ibid., 271.

9. Cheryl Misak, "Charles Sanders Peirce 1839–1914," *Cambridge Companion to Peirce* (2004).

10. William James, *Pragmatism: A New Name for Some Old Ways of Thinking: Together with Four Related Essays Selected from* The Meaning of Truth (1949), 50.

11. John Dewey, *The Quest for Certainty: A Study of the Relation of Knowledge and Action* (1929), 3–25, 254–286. Philip W. Jackson, "John Dewey," *A Companion to Pragmatism* (2006).

12. John Dewey, *Studies in Experimental Logic* (1916), 183–219. Jackson, "John Dewey."

13. Paul T. Durbin, "Pragmatism," *Encyclopedia of Science, Technology, and Ethics* (2005).

14. Dewey, *The Quest for Certainty*, 195–222, 254–286. Larry A. Hickman, *John Dewey's Pragmatic Technology* (1990, hereafter *JDPT*), 183.

15. See, for example, *Being and Time*, §§ 9, 13, 15. Heidegger's whole point of departure emphasizes the primordial unreflective, concrete, and—of course—temporal character of being in human existence. In this manner his project shares with pragmatism some ways of critiquing the traditional philosophy.

16. *Philosophical Investigations*, e.g., §§ 19, 23, 139, 206, 241. David G. Stern, *Wittgenstein's Philosophical Investigations: An Introduction* (2002), 160–170.

17. Karl Marx, "Theses on Feuerbach," *Selected Writings* (1994), 98–101. Frederick Copleston, *A History of Philosophy. Vol. 7: From the Post-Kantian Idealists to Marx, Kierkegaard, and Nietzsche* (1965), 305–334.

18. Joseph Margolis, *Reinventing Pragmatism: American Philosophy at the End of the Twentieth Century* (2002). Roger F. Gibson, Jr., "W. V. Quine," *A Companion to Pragmatism* (2006). Joseph M. Heath, "Jürgen Habermas," *A Companion to Pragmatism* (2006).

19. Carl Mitcham, *Thinking through Technology* (1994), 73.

20. Ihde, *Postphenomenology*. Mitcham, *Thinking through Technology*, 76–78. Verbeek, *What Things Do*, 99–120.

21. Mitcham, *Thinking through Technology*, 71–78. Hickman, *JDPT*, xi–xv, 1–16. Paul T. Durbin, "Introduction," *Techné* (2003).

22. Larry A. Hickman, *Philosophical Tools for Technological Culture: Putting Pragmatism to Work* (2001, hereafter *PTTC*), 12 (italics in the original). See also *JDPT*, 6–8. Briefly and roughly Dewey, in "What I Believe" (1930), characterized technology as follows: "Technology signifies all the intelligent techniques by which the energies of nature and man are directed and used in the satisfaction of human needs" (cited in *PTTC*, 8).

23. Hickman, *PTTC*, 54.

24. Hickman, *JDPT*, 40–41 (emphasis in the original).

25. Hickman, *PTTC*, 81 (italics mine). Dewey's notion of inquiry, in the briefest, equals "the general method of intelligence" (*PTTC*, 42).

26. Mitcham, *Thinking through Technology*, 74–75.

27. Hickman, *PTTC*, 18–19.

28. Hickman, *JDPT*, 21: "The purpose of the tool is to reorganize the experience in some way that will overcome its disparity, its incompatibility, or its inconsistency. . . . Tool is in this sense a theory, a proposal, a recommended method or course of action." See also *PTTC*, 12, 14, 71, 119.

29. "Habit" is another name for platform (Hickman, *JDPT*, 85). Technical tools can be both material and immaterial (*JDPT*, 46–48), and material tools always have an ingredient of intellectual meaning in them (*JDPT*, 21).

30. Hickman, *JDPT*, 19–23; *PTTC*, 18–19.

31. Hickman, *JDPT*, 70–71; *PTTC*, 21–23, 38–39.

32. Quoted in Hickman, *PTTC*, 39 (emphasis in the original).

33. Hickman, *PTTC*, 21–23, 38–39.

34. Hickman, *PTTC*, 37–39. See also Hickman, "Nature as Culture: John Dewey's Pragmatic Naturalism," *Environmental Pragmatism* (1996). Hickman claims that Dewey eschewed the notion of domination but embraced that of *management* concerning human relationship to nature.

35. John Dewey, *Logic: The Theory of Inquiry* (1939), 108. Hickman, *JDPT*, 45. (Italics in the original.)

36. Hickman, *PTTC*, 69–71.

37. This is somewhat surprising and confusing, since by and large Hickman argues that traditional philosophical problems are artificial as they arise from false dichotomies, or at least they can be solved in the pragmatist thinking.

38. Science and technology can be perceived within mere immediacy as well. The reception of science as an unproblematic body of knowledge and technology as neutral means represent examples of a not sufficiently mediated view of them. Hickman does not make this point in the mentioned context, but it is in tune with his general line of argument, although he does not write much about the problems of "scientism." See *PTTC*, 51, 72–75, 97; *JDPT*, 119.

39. Hickman, *PTTC*, 24–5, 83–88.

40. Ibid., 42.

41. Ibid., 43.

42. Hickman, *JDPT*, 31–32, 48–50; *PTTC*, 69–70.

43. Hickman, *PTTC*, 65–72.

44. Hickman (*JDPT*, 111) quotes C. Wright Mills's description of Dewey's method of recasting: "Dewey takes a point of sight and builds a conceptual structure within which he can grasp both the points which were being argued over; this structure is different from either of the conflicting or isolated doctrines which it 'combines.' It is Deweyan" (Mills, *Sociology and Pragmatism.* Oxford: Oxford University Press, 1966, 314. Quoted in Hickman.)

45. Hickman, *JDPT*, 99, 107–109, 112–114; *PTTC*, 7, 10–11, 31, 41, 109, 179–81, 183.

46. Hickman, *JDPT*, 10; *PTTC*, 39–40, 48; *Pragmatism as Post-Postmodernism: Lessons from John Dewey* (2007, hereafter *PPP*), 27–28, 116–118.

47. Hickman, *JDPT*, 23, 94, 96, 181–183, 196 (citing John McDermott). See also Keulartz et al, "Ethics in Technological Culture," 16.

48. Hickman, *JDPT*, 16, 21–22, 36, 40, 43, 46–48, 61, 173, 193, 202; *PTTC*, 12, 14, 17–19, 36, 46; *PPP*, 84–85, 121–124, 154–159, 166.

49. "Structures are, in Dewey's view, just 'constancy of means, of things used for consequences, not things taken by themselves or absolutely' (LW1:64–65; EN:62–63)" (Hickman, *JDPT*, 125).

50. Hickman, *JDPT*, 18–19, 23, 125, 129, 154–155; *PTTC*, 121–123; *PPP*, 123–125, 159.

51. Hickman, *PTTC*, 102. See also "Nature as Culture," 50–57. Hickman does not explicate to what extent this non-cognitive experience is conceptual or non-conceptual.

52. Hickman, *PPP*, 131–139; "Nature as Culture," 57–58. Nature-as-nature would be nature in modes of action 1 and 2, while nature-as-culture would be worked on in action 3 and reflected and reworked in action 4.

53. Hickman, *PTTC*, 54.
54. Ibid., 58.
55. Hickman, *JDPT*, 168–173.
56. Jonathan Glover, *Humanity: A Moral History of the Twentieth Century* (2001), 155–236. Langdon Winner (*Autonomous Technology: Technics-out-of-Control as a Theme in Political Thought*, 1977, 295–305) discusses the outbreak of the First World War under the theme of "technics-out-of-control." Both draw on Barbara W. Tuchman's *The Guns of August* (1962).
57. Hickman, *JDPT*, 175.
58. Hickman, *JDPT*, 173–175; *PTTC*, 55–56.
59. Hickman, *JDPT*, 168–173; *PTTC*, 57.
60. Hickman, *JDPT*, 169.
61. Hickman, *JDPT*, 73.
62. Hickman, *PTTC*, 66.
63. Hickman, *JDPT*, 12, 67, 90, 201; *PTTC*, 33, 72.
64. Hickman, *JDPT*, 33–34, 192; *PTTC*, 68, 106–108, 157–158.
65. Hickman, *JDPT*, 66–68.
66. Hickman, *JDPT*, 12–13, 148, 163, 169, 175, 201–202.
67. Hickman, *PTTC*, 161–164.
68. Ibid., 172–178.
69. Ibid., 164–170.
70. Hickman, *PTTC*, 170–172; *PPP*, 79–91. Cf. Andrew Feenberg, *Questioning Technology* (1999), 201–225. See below for Feenberg's response to Hickman's reading of him (Feenberg, "Pragmatism and Critical Theory," *Techné*, 2003).
71. Hickman, *PTTC*, 172.
72. Hickman, *JDPT*, 8–16, 140–165. Albert Borgmann, *Technology and the Character of Contemporary Life: A Philosophical Inquiry* (1984), 7–12.
73. Hickman, *JDPT*, 12–13, 148–149; *PTTC*, 72–73, 157–158; *PPP*, 67–68, 156.
74. Hickman, *JDPT*, 148–149.
75. Hickman, *PTTC*, 73.
76. Hickman, *JDPT*, 148–151. See also Langdon Winner, *The Whale and the Reactor* (1986), 3–18.
77. Winner, *Autonomous Technology*, 227. Hickman, *JDPT*, 152.
78. Winner, *Autonomous Technology*, 229. Hickman, *JDPT*, 152.
79. Winner, *Autonomous Technology*, 229.
80. Hickman, *JDPT*, 153.
81. Hickman, *JDPT*, 157; see also 156, 164, 170.
82. "Technological imperative" refers to chaining of technologies. Introduction of one technology easily entails imperatives for introducing others, both through new needs and through new possibilities. The introduction of electrical lighting, for instance, creates the need for power supply; and the power creates a possibility of introducing air conditioning. Hickman, *JDPT*, 151. "Autonomous technology" refers to the idea of technology having "a life of its own," that it develops independently of human intentions and plans. See Wha-Chul Son, "Autonomous Technology," *Encyclopedia of Science, Technology, and Ethics* (2005).
83. In Marx's work Hickman (*JDPT*, 142–146) distinguishes three different forms of "technological-stage determinism." These nuances are relevant but beyond the scope of this study.
84. In *PTTC* (115–128) Hickman criticizes "text type determinists," for instance Albert Borgmann (in *Technology and the Character of Contemporary Life* and *Holding On to Reality*) and Allan Bloom (*Closing of the American Mind*, 1987), who worry about other media having increasingly taken the place of the traditional book. Hickman's point, in brief, is that the reflective ingredients determine the level of art and thinking, not the medium employed.
85. Hickman, *JDPT*, 140–142.
86. Hickman, *JDPT*, 158–162, 120–131.
87. Hickman, *JDPT*, 145–146, 148, 160–162.
88. Hickman, *JDPT*, 161–162.
89. Hickman, *JDPT*, 154–158.
90. Hickman, *JDPT*, 162–165; *PTTC*, 36–41.

91. The classical distinctions concerning the three kinds of activities are found, for instance, in Aristotle, *Metaphysics*, 1025b17ff; 1064a5–6, 17ff; *Nicomachean Ethics*, 1139a26–b9; and *Magna Moralia*, 1197a9–13.

92. Hickman quotes Dewey: "There is no ground whatever upon which a logical line can be drawn between the operations and techniques of experimentation in the natural sciences and the same operations and techniques employed for distinctively practical ends. Nothing so fatal to science can be imagined as elimination of experimentation, and experimentation is a form of doing and making. Application of conceptions and hypotheses to existential matters through the medium of doing and making is an intrinsic constituent of scientific method (LW12:434–35; LTI:439)" (*JDPT*, 107). Dewey himself preferred the term "instrumentalism," which Hickman does not use much, since in the contemporary context of the philosophy of technology it has some evident wrong connotations.

93. Hickman, *JDPT*, 140–166; *PTTC*, 59, 72–76, 119–21, 125–28, 145, 157–58, 171. On the same points concerning pragmatism and technology in general, see Keulartz et al, "Ethics in Technological Culture," 6–7, 14–18.

94. According to Hickman (*PTTC*, 102), this view derives from the 16th and 17th centuries. Thus, I am not sure why he calls it the "Enlightenment" view. I guess Enlightenment here refers to the project of Enlightenment broadly, not to the movement (*Aufklärung, Lumières*) or era.

95. Hickman, *PTTC*, 100–104.

96. Hickman, *PTTC*, 88–90, 95–99; *PPP*, 48–62.

97. Hickman, *PTTC*, 104.

98. Hickman, *PTTC*, 100–104.

99. In plural, since Hickman (*PTTC*, 105) points out that there are many variants of fundamentalistic thinking. This theme is obviously vast and I will not go deep into it. Hickman touches upon Islam but those remarks are peripheral for this study. Moreover, the U.S. context of Hickman's writing explains the attention he gives to fundamentalism.

100. Hickman, *JDPT*, 33–34.

101. For some reason Hickman does not write about "intelligent design," which is a newer form of advancing "the creation science."

102. The selectivity of the biblical material in literal interpretations is a more well-known inconsistency. Some passages are read painstakingly literally, others are seen as irrelevant or outdated.

103. Hickman, *PTTC*, 105–108; *JDPT*, 191–192.

104. Hickman, *PTTC*, 100.

105. Hickman, *PTTC*, 108–111.

106. Hickman, *JDPT*, 100; *PTTC*, 113; *PPP*, 139.

107. Hickman, *JDPT*, 35, 81; 1996, 57–58; *PPP*, 148–149, 158.

108. Hickman, *JDPT*, 25.

109. Hickman, *JDPT*, 22. Hickman (*JDPT*, 27), moreover, notes that knowledge purely "by immediate apprehension" is an abstraction that does not take into account the conceptual productive activity that indirectly enables such apprehensions.

110. Hickman, *PTTC*, 113.

111. Hickman, *JDPT*, 99–105.

112. Hickman, "Nature as Culture," 57–58; *PPP*, 148–149.

113. Hickman, *JDPT*, 202.

114. Hickman, *JDPT*, 85, 98, 129, 155, 157, 164, 183; *PTTC*, 16, 23, 38–39, 132–133; *PPP*, 241–254. "Dewey argued that 'occupations' or 'vocations' (general patterns of productive activity to which an individual is committed and which are for him or her the locus of meaning) determine 'fundamental modes of activity,' and thus that they 'control the formation and use of habits' (MW2:41;PC:175)" (Hickman, *JDPT*, 85). "One of the impediments to meaningful choice most often mentioned by Dewey was the dead weight exercised by methods, habits, and institutions that have ceased to be properly instrumental, that is, instrumental to the ends for which they were chosen" (Hickman, *JDPT*, 157). "In sum, both Peirce and Dewey reached escape velocity with respect to the traditional and still-raging debate between the cognitivists

and praxicalists. They accomplished this by measuring human development and accomplishment in terms of neither ideas, nor activities, but in terms of habits produced. Their root metaphors went beyond theory and practice all the way to production" (Hickman, *PPP*, 254).

115. Hickman, *PTTC*, 132 (emphasis mine).

116. Hickman, *JDPT*, 154–158; *PTTC*, 38–39, 132–133. See also Hugh LaFollette, "Pragmatic Ethics," *The Blackwell Guide to Ethical Theory* (2000), 401–410.

117. Hickman, *JDPT*, 114. The Golden Rule is a principle of reciprocity most commonly shared by a variety of moral traditions. Its well-known formulation is Matthew 7:12: "In everything do to others as you would have them do to you; for this the law and prophets."

118. Hickman, *JDPT*, 188. Here one might want to protest that science and morals are different things and that testing and fallibilism applies only in science. This distinction, nevertheless, is exactly what Hickman's reading of pragmatism rejects.

119. Hickman, *JDPT*, 110–114; *PPP*, 185–187. "For Dewey, to be moral is to exercise efficient intelligence. On small scales and large, force is justified insofar as it serves the purposes of economy and efficiency; and it should be remembered that the law, which Dewey calls 'legal machinery,' has been set up as an 'energy saver'" (*JDPT*, 189).

120. Larry A. Hickman, "Making the Family Functional: The Case for Legalized Same-Sex Domestic Partnerships," *Philosophy of the Social Sciences* (1999).

121. Hickman, *JDPT*, 112–113. Hickman makes the argument in favor of consequentialism and cognitivism in ethics look easy. I do not, however, find his points very convincing. First, are impulses the only alternative for hypotheses and theories? How about moral intuition, conscience, sympathy, love, or compassion as elements in moral life? Second, Hickman does not provide many arguments for his consequentialism and against its obvious competitor, deontological ethics.

122. Hickman, *PTTC*, 147–156. This may sound like a dismissive perspective to begin with. Partly it is, as Hickman's reading of Ellul and Heidegger is not very detailed or charitable. Hickman, however, is looking at their projects on the whole and especially the most mythical and existential—religious in Ellul's case, quasi-religious in Heidegger's—elements in their understanding of the human condition in modernity. Of course, Ellul could be read as a sociologist of technology and Heidegger as a thinker who studies the Western metaphysical tradition. Both of them, nonetheless, professed more practical stances as well. Heidegger was most often elusive concerning them, but Ellul wrote practically oriented, hopeful books based on his religious views.

123. Hickman, *PTTC*, 151–153.

124. Ibid., 151–153.

125. Ibid., 149–151.

126. Ibid., 154.

127. Hickman, *JDPT*, 196–203; *PTTC*, 153.

128. Hickman, *PTTC*, 109–110.

129. Hickman, *PTTC*, 113.

130. "Technology" used here in its dictionary meaning, not in Hickman's broad sense.

131. "Popularization of science" carries connotations that make it an ill-fitted word in Hickman's context. "Public understanding of science" (Brian Wynne, "Public Understanding of Science," *Handbook of Science and Technology Studies*, 2001; Hickman does not employ the term) has become a customary way to refer to this topic. In the pragmatist context, and even generally, it could be modified into "public understanding of technoscience." "Scientific citizenship" has become an established term as well.

132. Hickman, *PTTC*, 113.

133. Ibid., 110.

134. Hickman, *PTTC*, 112. Dewey, MW.15.51.

135. Hickman, *PTTC*, 114. Dewey, LW.13.172.

136. Max Horkheimer, *Eclipse of Reason* (1992, first published 1947), 45–58.

137. Borgmann, "Kinds of Pragmatism."

138. Borgmann, "Kinds of Pragmatism," 30–31.

139. Larry A. Hickman, "Revisiting *Philosophical Tools for Technological Culture*," *Techné* (2003), 68.

140. See, e.g., Kristin Shrader-Frechette's *Environmental Justice* (2002) for dubious cases in technology policy. Of technological devices/systems consider, for example, if the assembly line truly provides room enough for human autonomy.

141. "Transcendent" in the sense that an other person is always transcendent for me; her inner life, for instance, remains not fully known or graspable with concepts. A person's spontaneity makes it impossible to objectify her fully. Absolute in the sense that Kant's (*The Grounding for the Metaphysics of Morals*, Ak. 428–429) formulation of the categorical imperative, which prohibits the merely instrumental use of a person, rests on fairly solid argumentative ground. Therefore, a thinker who would like to challenge it has an onus to provide arguments.

142. *Republic*, 502c–521c. Iris Murdoch, *Existentialists and Mystics: Writings on Philosophy and Literature* (1999), 299–385 (esp. 372–385).

143. Hickman, "Revisiting *Philosophical Tools for Technological Culture*."

144. As an aside: Jonathan Glover's *Humanity: A Moral History of the Twentieth Century* both documents and discusses the sinister character of the last century.

145. Feenberg, "Pragmatism and Critical Theory," 44–47.

146. Ibid., 46.

147. Feenberg, "Pragmatism and Critical Theory," 44. Hickman, *PTTC*, 81.

148. Hickman, *PTTC*, 120.

149. "And we can go through the many, many other groups of games in the same way; we can see how the similarities crop up and disappear. And the result of this examination is: we see a complicated network of similarities overlapping and criss-crossing: sometimes overall similarities, sometimes similarities in detail" (*Investigations* § 66). "Und so können wir durch die vielen, vielen anderen Gruppen von Spielen gehen. Ähnlichkeiten auftauchen und verschwinden sehen. Und das Ergebnis dieser Betrachtung lautet nun: Wir sehen ein kompliziertes Netz von Ähnlichkeiten, die einander übergreifen und kreuzen. Ähnlichkeiten im Großen und Kleinen" (*Untersuchungen* § 66).

150. Paul T. Thompson, "Putting Pragmatism to Work?" *Techné* (2003), 60–61. Borgmann, "Kinds of Pragmatism," 28.

Chapter Six

Thinking Further in Ethics and Technology

A great intellectual and practical transformation took place in Europe during the 16th and 17th centuries. Natural scientists, such as Nicolaus Copernicus (1473–1543), Galileo Galilei (1564–1642), and Johannes Kepler (1571–1630), rethought the methodology of science and made good practical use of the new ways of thinking.[1] This reform of science was not merely intellectual but involved employment of technical devices—telescopes and measuring apparatuses, for example—as means of doing experimental science. René Descartes (1596–1650) made central steps in modern mathematics by rejecting the ancient distinction between multitude and different kinds of magnitude, introducing new algebraic methods, and creating analytic geometry. In his philosophical thinking, Descartes was challenged by the crevice that had opened between immediate human perception and novel scientific discoveries. Descartes's philosophy came to lay down the fundamental set of assumptions and problems for the modern worldview.[2]

The moral and political thinking of Niccolo Machiavelli (1469–1527), Thomas Hobbes (1588–1679), and John Locke (1632–1702)—to mention three central figures—set a groundwork for a new human-centered morality, in which good and right were based on the judgments and agreements of individuals rather than the order of nature or divine legislation. This also meant "lowering the standards": thinking about human action was to take its bearings from actual life and its inevitable motivations, instead of following ideas of reason.[3]

In Francis Bacon's (1561–1626) thought, the new scientific methodology and the new ethos merged. Bacon emphasized that the new science was a pursuit in which nature was manipulated in order to acquire (scientific) knowledge, with which humans could (technically) conquer nature and make it serve their interests. This manipulative approach differed greatly from the ancient and medieval stances toward reality. The older stances emphasized reality—nature in itself or the world as God's creation—as a standard, that is, something to which humans were supposed to accommodate themselves, for example through practice of virtues and self-restraint.[4] In Bacon's approach the old model is reversed: human will and desire are granted the right and power to set requirements for reality.[5]

As we have seen, in addition to great improvements to the human condition, "the modern project" has produced a plethora of problems. Ecological problems that have resulted from technological advance are undeniable: thinning of the ozone layer, climate change, pollution, rapid extinction of species, and erosion. Pollution, moreover, is a real threat to human life.[6] Living with technoscience, as the previous chapters have pointed out, also involves certain moral, existential, and political problems.

Thinkers of the emerging modernity managed to transform the assumptions underlying science, production, politics, and morality in a way that made the new science, technology, and democratic forms of governance highly meaningful and desirable. Now we seem to be in need of a similar turning. Although debatable issues persist, it appears evident that humankind should alter the course that has been dominant in the modern era. If production and consumption are not modified in quantity and quality, the global ecosystem will suffer serious damage. Just as thinkers of the 16th and 17th centuries managed to make the modern project viable, some are now attempting to open the way toward an alternative modernity.

The need for an intellectual turn characterizes the calling underneath philosophy of technology. A number of other intellectual missions can obviously claim to be answering the same calling, but the thinkers discussed in the earlier chapters can legitimately argue to be among those facing the challenge. Hence it needs to be assessed, on a general plane, where and how far their answers can lead.

BEING AND GOOD

In societal discussions related to technology the concept of value is often invoked. Yet the problems of this notion became evident in chapter two, after a brief review of Heidegger's critique of it. Further on, we continued to encounter the unclear and ungrounded character of "values." If values merely depend on personal attribution, they become arbitrary; or alternatively, they are subordinated to techno-economic frameworks and do not include real moral and political reasoning that would be able to orient techno-economic development. The problem of values is a central theme in standard philosophy. Analytic philosophy approaches it via a metaethical analysis of realism against non-realism in ethics, value subjectivism against objectivism, and cognitivism against non-cognitivism. Continental philosophy places the question of values within a broad historical narrative: How did it happen that commanding Good turned into value based on an individual choice? The inquiry at hand aims to contribute to the understanding of this problem through studying values in the context of modern technoscience.

We saw how Hans Jonas treated the problem of value through questioning the modern notion of value-neutral being. Jonas argues that most branches of modern thinking—from hard sciences to existentialism—assume that reality in itself is devoid of value. This assumption creates the problem of value, since value is always something added to "the real" reality. "Reality" here may refer equally well to the 'nature' of the natural sciences or 'Being' in Heidegger's philosophy. In social sciences, moreover, even the human realm is framed for research as neutral entities. Jonas asserted that it is asymmetrical to claim that comprehensive thinking about nature is metaphysical while simultaneously holding on to the assumption of value-free nature, without presenting an argument in support of the assumption. It would be fairer to acknowledge that whether reality in some way includes good or not is a metaphysical question, and responding to it requires a metaphysical inquiry. Jonas's critique thus leads us under—as it were—the modern worldview and seeks the true Good in Being itself. The manner in which Being calls for response from humans reveals something of the Good to them. From the other direction, the capacity to have purposes that are inherent in all organic life is for Jonas a good-in-itself.

Jonas's critique of the modern value-neutral being appears cogent. It is asymmetrical to claim that connecting Being and Good is in need of an argument but that their separation can be taken for granted without any burden of proof. Further, studying what kind of demand Being issues to humans seems the correct next step. But here things also get nebulous. Jonas's arguments that take off from the analysis of organic life are interesting but not fully convincing.

Albert Borgmann's take on the concept of value rests on his characterization of the device paradigm. He maintains that thinking and talking about values mostly misses the point. Life in our technological era is structured in a certain way that evades analysis that relies on the human choice of values. To be able to ask the real questions, the structuring of our life needs to be made visible. The device paradigm is an ontological pattern of technology that structures contemporary life. It is not a direct result of conscious human planning or development. The device paradigm, nonetheless, crucially sets conditions for value discussions. "Values" appear to fall either on the foreground or the background side of the device paradigm. As "foreground" decisions they are easy, light, and arbitrary surface choices of things one "prefers" or "does not prefer." This kind of light value talk is problematic, since within it values seem to lack a binding force and reasonable conversation about them seems to be out of place. If values can be chosen, how can they have commanding power? As elements of the background of technology, values are sturdier. In the background, however, the possibilities for choice appear small. Little room exists for realizing, for instance, "environmental values." If some room is made, it is translatable into the amount of money sacrificed. As values are functions of techno-economic boundary conditions, they are governed by the device paradigm. In this sense value talk stays within the pattern of technology.

Borgmann's own solution for the lack of depth and orientation in life is engagement with focal things and practices. In the context of things and practices, the value concept, which is based on the distinction of means and ends, is not really applicable. Living with things is more concrete, richer in context: they anchor a person, bring one to a reality deeper than the modern framework of subject, object, choice, and value.

Borgmann's careful analysis of the structuring force of technology is necessary for an appropriate discussion of morality and the good life in the contemporary world. His analysis, however, does not prove the concept of value to be wholly lacking in viability. One could argue that such issues as human rights, distribution of wealth, capital punishment, and economic policies related to distribution of wealth, for example, are genuine moral and political questions that could also be called value questions. This, in fact, coheres with Borgmann's conclusion that the device paradigm has its place in the infrastructure of the liberal democratic society.

Hickman has no reservations about using the concept of value. Yet he uses it in a particular pragmatist sense. For Hickman values are functions of adaptability. One tool or habit is more valuable than another—in case the comparison is relevant—if it enhances human adaptation to the environment more than the other. Inquiry is the method to assess values. Conservative Christians, for example, make a value argument that gay couples are not good parents for children. According to Hickman's reading of the research data on the topic, this argument is based on old prejudices and dysfunctional values, since social scientific studies do not support it. Therefore, he argues, it can be rejected as false. Not all old moral traditions, however, are in need of revision. Hickman mentions the Golden Rule as a moral tool that can stand the test of inquiry. As pointed out in chapter five, however, it is not clear what is the criterion or criteria against which the Golden Rule—or any other moral notion—was measured when it passed such a test.

To an extent, applying the experimental method to moral questions is, of course, a good idea. According to the mainstream view, moral argumentation consists of moral elements—values, norms, and clarifications of moral terms—and factual elements: facts and conditional clauses concerning the future ("if we do x, y will follow"). Insofar as this is the case, factual elements are fully open to experimental revisions. Yet Hickman does not accept the fact/value distinction. His arguments against the distinction are strong. But it is somewhat unclear how thinking and experimenting on morals in practice take place without using the fact-value distinction—especially when it is not clear how the meta-criteria for "moral tools" work.

Jonas and Borgmann agree that the proper way to think about *real ethics* is to question the modern, not fully articulated ontology and go beneath its framework. Hickman does not search for a source of values. His solution emphasizes the relational character of values, the way they depend on valuation, habits, and knowledge about their functionality.

One more way to frame the issue of science and value is as Max Weber did in his 1918 lecture, "Science as a Vocation (*Wissenschaft als Beruf*)." Weber formulates the problem as follows:

> Under these internal presuppositions, what is the meaning of science as a vocation, now after all these former illusions, the 'way to true being,' the 'way to true art,' the 'way to true nature,' the 'way to true God,' the 'way to true happiness,' have been dispelled? Tolstoi has given the simplest answer, with the words: 'Science is meaningless because it gives no answer to our question, the only question important for us: "What shall we do and how shall we live?"' That science does not give an answer to this is indisputable. The only question that remains is the sense in which science gives 'no' answer, and whether or not science might yet be of some use to those who put the question correctly.[7]

Weber's own treatment of the question unfolds as ambivalent. On the one hand, he perceives the disenchantment brought by science as inevitable. No "re-enchantment" could be possible, since science has shown the illusoriness of most earlier positions and worldviews. On the other hand, he longs for meaning, value, and depth that appear absent and impossible under the conditions of disenchantment. In the modern era, according to Weber, science is the destiny of humankind. Science does not, however, address the meaning of being, does not reflect on true happiness; it brackets value and is oblivious of death as the dreadful marker of human finitude.[8]

In addition to Weber's resignation in the face of scientific value-emptiness, two sources of meaning appear as options. First, an existential stance toward the world and one's own being could provide life with meaning. As Weber notes, one must choose what gods one serves. In a word, this source of meaning boils down to "faith," in an inner, subjective sense. Second, another source of meaning could be notions that go beyond the scientific worldview, e.g., ideology or metaphysics, that would grant being some value and meaning. A metaphysical interpretation of scientific facts or a system parallel to science could furnish the world with purpose, meaning, and maybe transcendence. Both options, nonetheless, are problematic. As for faith as an existential stance, in Heidegger's case we noticed that it is difficult to find real criteria for evaluating the content of such a purely inner faith. It is either empty in content or dependent on system of thought. Of a purely formal and self-referring faith Heidegger's resoluteness (*Entschlossenheit*) in the 1920s serves as an example. Many a faith may give life a thicker meaning, but as a contentless faith seems impossible, faith usually needs ideological backing. In his personal life the self-referring of Heidegger's resoluteness led him to a somewhat unsophisticated ideological commitment in the early 1930s.[9] The problem of ideologies is twofold. First, in the modern world everything that is not strictly science based is somewhat arbitrary. Consider homeopathy, psychoanalysis, and Jonas's or Alfred North Whitehead's metaphysics. Hence, in light of science and the culture of reason most, if not all, ideologies appear arbitrary, if not flat wrong. Second, ideologies threaten the freedom, authenticity, and "freshness" of existential life. In sum, these tensions appear to take place within a triangle, like the one shown in figure 6.1.

Science possesses validity of a high degree and has become the measure of the real in the modern world. But science is not existentially life-giving because it is unable to answer questions concerning meaning, good, and obligation. Existence in its subjectivity is its own foundation and criterion. It is, exactly for this reason, capable of untruth and morally bad.[10] Because science does not provide meaning and because subjective existence is lost, without orienting marks, comprehensive exposition answering the big human questions is needed. Ideologies, however, either lack scientific support or are

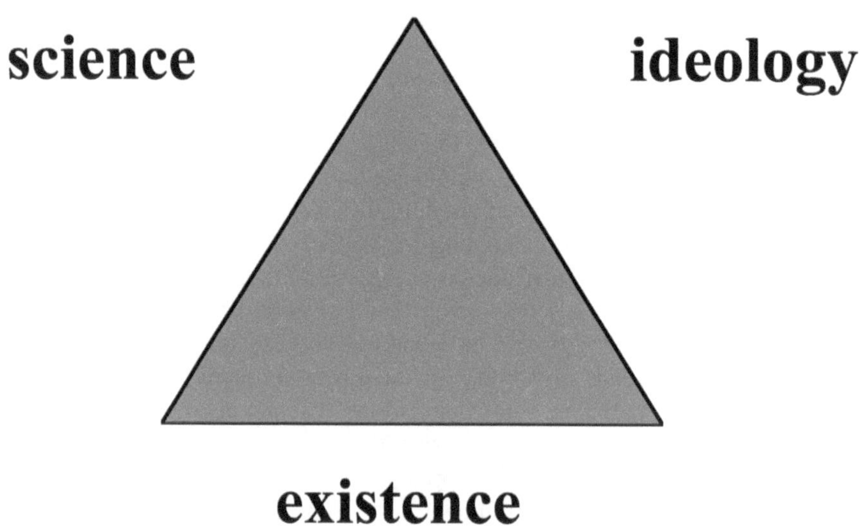

Figure 6.1.

in tension with science. "Doctrines," moreover, are often unpalatable for the modern person, since they suppress authenticity, the freshness of one's ownmost being.

Jonas's way to tackle Weber's problem is to emphasize the dialogue between science and metaphysics (ideology). Jonas thinks that science inevitably rests on some metaphysical/ideological presuppositions that need to be scrutinized. The value neutrality of being, dualism, and the implicit assumption that dead things are more understandable than the living are such assumptions. Jonas, however, wants to give science its due, and his thinking is in true dialogue with the natural sciences—unlike many other phenomenological thinkers.

Borgmann's philosophy pays attention not especially to intellectual constructions or to faith as an existential stance—although it does not exclude them—but to material culture. Borgmann perceives the emptiness opened up by science and technology as a possible condition for concrete things to come to the fore. Things are beyond existence, science, and ideology. In a technological world the answer to Tolstoy's question "What are we to do?" cannot be a mere thought or piece of language. Both questioning and finding answers require a context in which to take place, Borgmann argues. The context is temporal—a practice—and material—composed of things. Borgmann's approach, therefore, primarily addresses the material conditions of the good and meaningful life. At least to a certain extent his ideas are applicable independent of one's ideological commitments.

Hickman's pragmatism expands the notion of science. Hickman's technology, which equals Dewey's inquiry, should be able to address existential questions and assess ideological assumptions. Hickman thus argues for a broadly understood naturalism: all notions ought to be submitted to research processes in the laboratory of life.

Of the three approaches, only Borgmann's manages to find a way out of Weber's problem. As we saw earlier, in Jonas's approach, metaphysics, which to an extent escapes both philosophical and scientific critique, has the last word, and in Hickman's, ethics becomes thin and space for existential thinking small, as in the end (pragmatistically conceived) science has the last word. Borgmann's thinking, to the contrary, does not get entangled in the problematic region described by Weber. His thinking on material culture and things takes us outside the problematic relationships of science, existence, and ideology. Moreover, Borgmann manages to give all three their due better than Jonas and Hickman, because he does not favor any of them one-sidedly.[11]

TOWARD AN ETHICS OF TECHNOLOGY

This inquiry has observed several distinct strokes in the frontier of comprehensive thinking on technology and ethics. To have a coherent picture of the matter, these strokes need to be brought together. I will try to do that through four themes. First, I will revisit the distinctions between making, doing, and knowing. Then I will sketch a vision of mid-level ethics that will frame a central domain for ethics of technology. Third, I will present a minimum task for ethics of technology—mission that at the very least should be carried on, no matter how much the present ideals of neutral research curb thinking. Finally, in contrast to the minimum task, I finish with suggesting a more ambitious maximal ideal for ethics of technology.

On the Moral Characteristics of Technology

Distinctions between science, technological production, and moral thinking have been criticized throughout this inquiry. For the most part, I have commented approvingly on these criticisms. Some clarifying and concluding remarks on the topic, however, are due. First of all, I do not think any good arguments for the neutrality of making, poiesis, or technological production exist. I find Langdon Winner's account of reverse adaptation a convincing description of many phenomena in contemporary society: "Reverse adaptation—the adjustment of human ends to match the character of available means."[12] Below, the brief description of the fetal screening practices serves also to shed light on reverse adaptation. These practices cannot really help the fetus, since essentially the only positive action that can be taken if something is found to be wrong is an abortion. Still, the screenings are understood as "care" for the fetus.

Further, as I concluded in chapter four, I think Borgmann's analysis of the contemporary culture in light of the device paradigm is, for the most part, valid. This means that human engagement with technological devices and systems affects the whole way humans exist in the world. Hickman's account of habituation agrees with many of Borgmann's points, although Hickman does not want to think that the structuring power of technology is as thorough-going and all-pervading as Borgmann's theory argues.

But what is the relationship of moral reflection and action (praxis) to production? To begin, non-neutrality of poiesis does not abolish the distinction between praxis and poiesis. Material culture is not morally neutral, but moral characteristics within material culture are different from morality within praxis. Within praxis human beings practice self-conscious moral deliberation, but the products of poiesis do not. It seems clear that praxis is separate and independent of poiesis. Moral governance of poiesis, deliberation on design for instance, is best conceived as subordinate to praxis. It follows that if we have become aware that technology policy, designing technological artifacts, and using them are morally significant activities, they are in need of practical review. I find Hickman's account to blur the distinctions between praxis, poiesis, and theory too much. It seems that the three categories are worth keeping separate but in need of serious revision.

Furthermore, I argue that praxis is the domain in which the first human questions take place. Tolstoy's question "What shall we do and how shall we live?" is the first and most important philosophical question. For in human life acting is inevitable: humans cannot help but live and to live means to act constantly. For conscious beings acting means choosing between alternate courses of action, and talking beings give verbal accounts of their choices. Therefore, questions concerning proper and right action as well as its good ends are the first questions. In a sense, even thinking is action. This is revealed, for instance, by the English phrase "to pay attention to something," which indicates that humans choose where they direct their attention and are aware of "the costs" of choice—not being able to put attention somewhere else. Hickman elaborates well on this phenomenon by noting that "facts" are always chosen and that they get their meaning from being facts of a case. This position could be called "ethics as the first philosophy." It is an ancient position, not quite as old as philosophy itself, but emerged with Socrates.[13] Socrates used to abort an inquiry if he saw that it was not going to contribute to ethical questions that he perceived the most important.[14]

Of the three categories, the role of science turns out to be the most complicated. Science is problematic in many respects. On the one hand, science holds an unquestioned hegemony in description of reality. On the other hand, science penetrates our societies in its instrumental role but science as a true description of the world fails to reach large parts of society. Thus, at least as a social phenomenon, institutional science has a problem of communicating to the public. Even on a more epistemological level, this inquiry has reviewed a number of convincing arguments against the position that science is neutral realistic description of reality. First, modern science most often is highly dependent on technology—for example, in abstraction and manipulation of research objects as well as making them perceivable. From this it follows that, at the very least, claims about science as realistic description of reality should be qualified. Second, due to practical institutional arrangements and funding, science has an increasingly strong instrumental role in society. Although this does not mean that conceptually, science as truth-seeking activity would be impossible, science existing in reality is less and less a purely truth-oriented project. This is the case in both science as a social institution and the personal aspirations of scientific researchers. For researchers science does not seem to be a vocation or calling but a career in prudential terms. Third, social studies of science show science as human action and as a craft. These observations point at features in science that clearly are value laden and in need of moral and political evaluation. Fourth, it appears evident that a great deal of valuing—directing attention, choosing paths, allocating time and resources—takes place before doing science—thinking, experimenting, calculating.[15] In sum, modern science includes many value-laden aspects. These need to be assessed morally and politically. Sometimes contemporary science is ambiguously represented in terms of earlier forms of theoretic activities that had more justified claims for being morally independent.

The above points do not mean that in principle science could not be independent of practical interests. In this sense Borgmann's point that science should be kept conceptually separate from technology is valid. As universities face more and more demands on efficiency and productivity, disinterested basic research is pushed to the margins. From a purely philosophical point of view academic culture seems to need a revival of theoretical life—putting more emphasis on science as search for truth. Yet in practice such a revival appears unlikely to take place.

Mid-level Ethics

As has become evident at several points, applied ethics emerged on thin philosophical foundations for normative moral thinking. Had the historical contingencies not been what they were, closer contact with the Western tradition of ethics and political philosophy as well as modern social sciences would most probably have benefitted the nascent applied ethics. Some development in these directions has, in fact, taken place more recently.[16] After the phase of non-practical metaethical analysis and "the practical turn" in the form of modern applied ethics, Anglophone ethical discourse may be interpreted as at present being involved in "an institutional turn." This means paying more attention to practices, praxes, institutional frameworks, and policies. While earlier applied ethics looked at how ethical agents use their agency in given situations, the institutional turn requires study of the formation of agency as well.[17] The observations and conclusions in chapter 2 pointed at this direction. Consider, for instance, the limitations noticed in the frameworks of engineering ethics and bioethics.

All the approaches examined in this inquiry suggest a need for an institutional turn—Borgmann's and Hickman's in particular. Borgmann's work sheds light on how material culture shapes the conditions for pursuit of the good life.[18] He argues that it is somewhat misplaced to discuss rational choices, if the context for those choices already—with or without awareness of this—favors certain courses of actions. For instance, having personally deep, and thus morally significant, communication within a family has little chance to occur if place, time, and appropriate things for it are missing. Consequently, Borgmann emphasizes moral reflection on things, practices, material culture, and places of communal life. His emphasis on things forms his most unique contribution to ethics. Mainstream ethical thinking—correctly—is the most concerned about the relationships between persons. Yet even if we maintain that personal relations are the most important subject for ethics, we still need to notice that people almost always do something together—exist within praxes—and that their doings have a material context. These secondary and tertiary subjects affect the primary themes—human beings' relationships to others and to themselves.

In this respect Hickman agrees with Borgmann. Hickman emphasizes how "an individual" is an abstraction, as humans in reality are embedded in their concrete and practical environments. These environments have a role in shaping human desires and subjectivity, and hence need to be taken into account when human freedom is considered. For ethics, in Hickman's work the most important locus is reflection and experimentation on habitual actions—habits or practices within which reflection usually lies dormant.

> Habits bind us to old ways of doing things and make us reluctant to initiate novel courses of action. But they, and especially the shared habits we call 'institutions,' also provide the springboard from which new discoveries can be made and new paths of action can be entered upon. Habit and thinking are thus for Dewey not polar opposites, but phases within human experience. Habit determines the channels within which thought operates, and thinking takes place within the fabric of habits.[19]

Moral reflection on habits, practices, and institutions takes place in "the mid-level."[20] It resides above the micro-level that looks at an isolated act by an individual, the common level in modern moral reflection. Yet it takes place below the macro-level that discusses the ultimate and most general principles of justice, public good, and social contract. This mid-level, or "meso-level," ethics looks at policies: individual, in group contexts, communal, and societal. Mid-level ethics is concerned with how life is arranged. As with many other aspects of human affairs, these arrangements were treated in detail in ancient philosophy: Plato, Aristotle, Stoics, and Epicureans discussed various ways of conducting life (*bios*).[21] Their conceptions of moral philosophy were at home in the mid-level, unlike modern ethics. The modern technologized context, however, requires new astuteness in the analyses in the mid-level, as Borgmann's work most clearly shows. The revival of mid-level, of course, has been on the way for some time. Forms of virtue ethics such as Alasdair MacIntyre's as well as Michel Foucault's (1926–1984) work are probably the strongest—but quite different—currents in the revival of the viewpoint focusing on practices.[22]

In the ethics of technology this emphasis on policy-ethics means that the mid-level, or meso-level, is often the most relevant sphere within which to ethically assess technological developments, artifacts, and structures. At the same time, it compels an encounter with culturally touchy questions that can be evaded when delegated to the choosing individual or diluted into macro problems that mostly concern legislation and the distribution of wealth. Consider, for instance, technological practices in fetal screening. New technological means such as genetic amniocentesis, ultrasound scans, and tests for the mother's blood markers enable doctors to gain information about a fetus's condition early on. Due to these technological possibilities for prenatal screening, the number of disabled babies being born in advanced countries is decreasing. Births of babies with Down syndrome, for instance, have decreased.[23] This is a policy that takes place. The policy, like most policies, is a result of many contingent historical factors. These factors include economic calculations on health care, various stages of the national health policies in the 20th and 21st centuries—including the eugenics (German: *Rassenhygiene*) of the era before the Second World War, the natural desire of the parents to have healthy children, and concerns articulated by the associations for disabled people. On the general level, however, the question: "Do we want fewer people with Down syndrome to be born?" has not been consciously addressed in a public discourse.[24]

In Finland, for instance, concerning this question the micro-, meso-, and macro-levels appear to be in contradiction. The official national public health policy and the policy of disability claim that all the citizens have an equal right to existence and life. Moreover, according to the explicit statements, the society promises to take care of its weaker members. Simultaneously, the fetal screening policy appears to be a tool for decreasing disabilities. This goal is not articulated explicitly, although calculations on cost efficiency are made. The official purpose of the screening is to increase "the reproductive autonomy" of the parents—especially of the mother. As this study has indicated, "autonomy" is not an unproblematic concept. Further, it may be that the mother primarily does not want more freedom but wants to help her child. In any case, it is unclear how screening increases autonomy, as the parents are seldom truly aware that the primary ultrasound tests are in fact voluntary.[25] Even further away is the awareness that the parents are participating in a screening process. The ultrasound tests are most often done under the guise of being treatments for the fetus, although no real treatment other than abortion is available, if the fetus is found to be disabled. Thus the tests in fact set conditions for the acceptability of the fetus's birth. Further, it is considerably difficult to opt out of the screening procedures if something worrisome is found. Finally, studies indicate that the health care personnel often express clear opinions about the correct choice for the parents who are expecting a disabled baby—not aborting the pregnancy is considered an awkward choice.[26]

This example can be analyzed in the categories of macro, meso, and micro. On the macro-level there is the principle of equal rights for all people regardless of disability as well as macroeconomic considerations on health care costs. On the meso-level there is the screening practice that, according to studies, follows its own patterns that are largely unexpressed. On the micro-level there is the idea of the autonomous subject making choices.[27] Between these levels considerable gaps and even contradictions exist. The meso-level, nevertheless, turns out to be the decisive level, since on it the practical and material goings-on are settled, while micro- and macro-levels contain some optimistic but unrealized assumptions. In a cynical sense, this state of affairs appears to be quite functional: high ideals are affirmed, citizens' freedom increased, and health care costs cut.

I conclude that both the lacunae in theoretical work in ethics and the tensions in concrete practices point out the relevance of studies on the mid-level. Morally important inquiries may take place on a level that is somewhat indirectly ethical. Hence the relevance of STS for ethics of technology. Social scientific studies on technologies and technologized parts of the human world are able to provide knowledge without which ethics in the technological world would be inadequate. Admittedly, ethics becoming dependent on an increasing amount of external information might trouble an ethicist, as many towering figures—for instance, Socrates, Plato, the Stoics, Kant, and Kierkegaard—in the tradition of ethics have emphasized that right conduct and the good life are not and should not be dependent on such information. The need for relevant information, however, is one of the central aspects of the modern condition. It obviously behooves ethics as an academic discipline to take this into account. Hickman's classification of human action is helpful here. In reflection on morality appropriate knowledge is surely needed, but it may well be that more habitual action can still take place without a burden of knowing much—if the platforms for such action are well thought through.

The meso-level, nonetheless, has been discovered by management as well. Frederick Winslow Taylor's scientific management in the late 19th century presented an intervention in the meso-level. Taylor's idea was to employ a scientific method to improve the production processes: to study the working habits and techniques to enhance them according to empirical knowledge concerning the processes. Since Taylor we have seen different ways to control work efficiency. In the early 21st century this control increasingly emphasizes "quality" in addition to, or instead of, simple efficiency. This kind of intervention raises questions about the fate of human autonomy and dignity in such an instrumental approach.[28] Therefore, meso-level ethics should also include critical analysis of the attempts to improve and enhance practices and institutions.

The need for mid-level ethics elaborates further why interdisciplinary elements are necessary in the ethics of technology. In studying practices, institutions, and policies ethical assistance from social studies is needed. Sometimes just becoming aware of the broader social character of one's surroundings can have indirect moral relevance. Research ethics, for instance, is often thought of within a narrow applied ethics framework as good scientific conduct.[29] Good conduct is obviously important, but making students and researchers aware of the character of the research institutions within which they work as well as wider science policy can have deeply moral implications. For example, graduate students in natural sciences often see themselves as "doing basic research," i.e., aiming at broadening and improving the body of scientific knowledge, acquiring knowledge for the sake of a truer picture of the world. If that is the case, the ethical considerations related to their work are quite different from those related to research that aims at technological applications. Often the graduate students are not much aware of their place in their research group, in the academic institution, and in the whole technoscientific system that national science and technology policies aim to steer. It is not uncommon that a student who thought she was doing basic research only later finds out that her research group, in fact, has sold several patents.[30] The students could be made aware of this larger context of their research work by explaining the *de facto* conditions in their particular case, which is something that the research ethics codes ought to require the employer to do. But a more academic way to do this would be to include relevant STS literature in research ethics classes. This would enable the students to see the role of their work on a more general level and reflect on the place of science in the contemporary world. This is something one would expect from a person working on her PhD. Achieving this broader awareness appears necessary in order to indirectly provide possibilities for proper research ethics. This need as well as its counterpart—the request that scientists communicate their results better to the public—is strongly accounted for by Hickman.

In sum, looking at mid-level is necessary because real human action is seldom so isolated and individualistic as mainstream applied ethics assumes. Secondly, focusing on instantaneous decisions neglects the temporality of human life, in which repetition, practices, habits, and unreflected skills are central. These, to a great extent, shape human agency in practical action. Thirdly, narrow intellectual approaches miss the role of material culture in ethics. And material culture is mostly analyzed in mid-level. Finally, to repeat a point already made, in applied ethics technology is most often understood instrumentally, as a transparent tool. To amend this naïve view of technology, applied ethics needs to learn from the philosophy of technology as well as science and technology studies.

Minimum Task for Ethics of Technology

Given the context of liberal democracy and the scientific research community, the task of ethics and the ethicist is in certain ways limited. In democracy, moral expert is an awkward if not a nonexistent role, since the final decisions on moral matters belong to the public or to each citizen. The ideals of science curb the normative leanings of ethical theorizing to a large extent. Within these limits "a minimum task for the ethics of technology" can, nonetheless, be envisioned. This minimum task is to reveal where ethical questions have been ignored, suppressed, or covered over by technical questions. After pointing out ethical questions, ethicists, moreover, should point out ways to address them publicly. Yet it might happen that the public needs the ethicist's help in addressing moral issues related to technology. Offering such help appears to go beyond "the minimum task" sketched here and points toward a possible wider task.

An obvious domain in which ethics of technology can have an impact is education. Engineering ethics, research ethics, bioethics, and business ethics are taught as part of undergraduate and graduate programs. As thinking in ethics of technology progresses it will become broader and politically more relevant and influence the textbooks and curricula of the ethics classes. Hickman's approach to ethics raises the question of whether separate ethics classes are in fact needed or whether ethics should penetrate the whole curriculum and whether this kind of overall moral awareness would be sufficient. Constructing curricula so that they enable ethical thinking is of course an excellent idea. But as ethical reflection has a certain independent role in human life it should also have an independent place in education.

Related to ethics education, minimally conceived ethics certainly has a role in thinking about professional codes of ethics. Professional ethics may strive for more substantial stances, beyond mere legality and contractarianism, through reflecting on the character of the profession—foremostly types of engineering. Firstly, a profession aims at producing a common good beyond the mere self-interest of its members. For instance, physicians look at their work in light of life and health. Recently, engineers have begun to frame their profession as striving for public safety, health, and well-being. Secondly, profession as a practice contains principles concerning goodness and excellence within the profession. Hence, by thinking through the professional practice, professional virtues, i.e., the character of an excellent professional, can be articulated. This sketch of professional ethics, however, reveals the shortcomings of the minimally conceived ethics. What "safety," "health," and "well-being" consist of is obviously a huge ethical and political question. Moreover, if professional ethics wants to avoid being relativistic it needs to be able to see beyond the contingent practices and to assess the goodness and excellence that the professional practice rests on.

Furthermore, parliamentary technology assessment is a domain in which ethicists can be active and taken into account. Technology assessment is practiced in most advanced countries in some form or another. Prominent technology assessment institutions include, for example, the Office of Technology Assessment (in the United States from 1972 to 1995), the Rathenau Institute in the Netherlands (founded 1986), the Japan Science and Technology Agency (JST; preceding organizations started at the turn of the1950s and '60s), and the Danish Board of Technology (*Teknologirådet*, founded 1986). Technology assessment, in the best cases, influences legislation, policy choices, and funding decisions, and therefore can have a real effect on technological development. Information acquired from experts and democratic participation are not sufficient elements for decision-making in technology assessment; moral reflection is needed as well. The contribution of an ethicist can here consist of clarification of arguments and outlining the central moral positions in the given case, which together enable better democratic decisions.

Morally and politically, critical thinking can of course engage in existing processes actively, without an invitation from the legislators and administrators. In the spirit of mid-level ethics, ethics of technology could combine methods of policy studies and study through a moral-political lens what is going on in the mid-level of different policies—in research, environmental protection, design of public spaces, and education.

In the main, the minimum reading of the task for ethics of technology is procedural. This is due to the contexts of democratic governance, science, education, and free market, which do not allow much in the way of substantial stances. Minimally conceived ethics of technology does not make many substantial statements but points out inconsistencies and ambiguities and indicates ways for better education, democracy, and decision-making.

Maximal Ideal for Ethics of Technology

This inquiry has at times come across proposals for reform in the technologized society. Although advocating social change is peripheral at the most amongst the goals of contemporary academia, it is worth asking: How likely would such a reform take place? One cannot avoid thinking that many proposals reviewed in this inquiry appear unrealistic. Yet Western societies have witnessed forceful, transforming social movements during recent decades. Feminism and the environmental movement are two clear examples. The labor movement's successful efforts at improving social justice in capitalist societies formed an earlier instance of a realized reform. Critical thinking on technology, in fact, has contributed to environmental awareness. A comprehensive "technological consciousness," however, still stays far from reality.[31] Technology has not become political to the same extent that the environment did in the 1960s and '70s. The consequences of technology are usually discussed in the context of environmental policy. Further, science and technology policy addresses research and development as instruments of economic growth and as challenges for education. Technology itself, nonetheless, remains outside these considerations.

The concern for the environment has also a character of not being truly engaging and meaningful for individuals or political institutions. Views about what should be done are quite far from what is actually done—with the related guilt and bad conscience. I suggest that this is at least partly because we have not managed to bring the deepest issues under scrutiny. A real transformation in our basic assumptions about life, knowing, production, and economy might make environmental action truly viable. Human action that has most drastically transformed nature is, in fact, technological action. It is therefore probable that better and more intense thinking on technology would enable more motivated and properly oriented environmental action. I think Borgmann's analysis of the device paradigm showed well how thinking about technology itself can lead to the possibility of deeper environmental awareness in a meaningful and guilt-free way.

In philosophical work hypothetic and daring proposals are possible. If ethics of technology should be cast in stronger terms, how would it look? What would be "a maximal ideal" for ethics of technology? What would be the more substantial issues to be addressed? I think the most important of the little-addressed themes in philosophy of technology is the will, or volitional stance, underlying our technoscientific modernity.[32]

Bacon and Descartes, thinkers who centrally set the groundwork for modern technoscience, explicitly let this project hinge on certain volitional stances. In *Novum Organum* I.129, Bacon bases his approach on ambition:

> Further it will not be amiss to distinguish the three kinds and as it were grades of ambition in mankind. The first is of those who of desire to extend their own power in their native country, which kind is vulgar and degenerate. The second is of those who labor to extend the power of their country and its dominion among men. This certainly has more dignity, though not less covetousness. But if a man endeavor to establish and extend the power and dominion of the human race itself over the universe, his ambition (if ambition it can be called) is without doubt both a more wholesome thing and more noble than the other two. Now the empire of man over things depends wholly on the arts and sciences. For we cannot command nature except by obeying her.[33]

It is admittedly in many ways debatable to what extent ethos propagated by Bacon has ever underlain the modern technoscientific project, if at all. Moreover, Descartes's account of mastery of nature rests on the virtue of generosity (*generosité*), which he discusses in *The Passions of the Soul* (esp. articles 50 and 147–161).[34] This account is complex but the core of generosity is the resolution to make best use of the free will. The opposite of generosity is to be at the mercy of passions, chance, and outside circumstances. Without going into many details, I will make three remarks. First, in this study I have taken Bacon and Descartes's promises as centrally having been fulfilled. I will thus continue following that line of argument. Second, both "ambition" and "generosity" can without artificiality be read as affirmations of the will. Third, for ancient thinkers ambition was problematic even as fuel for political life, which can easily be seen, for example, in Plutarch's descriptions of Caesar and Antony. Aristotle, moreover, underscores how political life is dependent on honor and therefore on virtue; he counts straightforward ambition for power as lawlessness.

In this study, the problem of will surfaced most acutely in the chapter on Jonas. Looking at life, Jonas finds value-positing life as the grounding phenomenon for ethics. Looking at the same life, Arthur Schopenhauer (1788–1860) saw blind will to live, the value-positing of which is to a great extent destructive. Schopenhauer's pessimism is in the thesis: all being is willing and willing is destructive.[35] Strife and destruction result especially from the willing beings' reliance on the principle of individuation, separate identity in space and time. Consequently, for Schopenhauer the ethical task is to deny the will through compassionate action and *askesis*. This task relies on the metaphysical insight into the mere phenomenality of individuation: compassion (*Mitleid*)—feeling another's suffering or "suffering together"—is based on the knowledge of the unity in noumenal reality.[36]

For Schopenhauer, will posits values but the real ethics is found in the denial of the will and the value-positing subject. For Jonas, the basis of ethics is found in the organic reality of purposefulness and affirming the will to live in life. In their pessimism and optimism Schopenhauer and Jonas are almost diametrically opposite. Schopenhauer's compassion and Jonas's responsibility, nonetheless, seem to have much in common. The basis of compassion, for Schopenhauer, lies in understanding the illusoriness of the separation of two beings. The basis of responsibility, for Jonas, lies in understanding the perishability of all beings, myself and those beings for which I feel myself responsible. In Jonas's thinking, however, preserving the separate organic identities of beings is crucial.[37]

Willing in technology is most relevantly studied in a more limited and historical sense. Carl Mitcham summarizes:

> Speculatively and impressionistically, . . . it is perhaps permissible to suggest that the pursuit of efficiency and the will to control might even be termed a historically unique volition that can be associated with technology in a new way.[38]

Mitcham's characterization fits such varied thinkers as Martin Heidegger, C. S. Lewis (1898–1963), and G. H. von Wright (1916–2003), who perceived the manipulative ethos, or will to power, as inherent in modern technoscience. Further, they found this manipulative tendency, striving for mastery of nature, to contribute centrally to the problems into which technoscience has led our civilization.[39]

Most of the earlier philosophical takes on the will in modern technology have to a great extent remained reliant on the genius of the thinker who articulated them. Yet somehow the willing underlying technological activities—individual, institutional, and global—should be brought under rigorous collective scrutiny: under an inquiry that could offer itself as a theme of an intersubjective project. I find this to be the key challenge against which Hickman's techno-logy should be tested: How can our collective reflection and experimentation further criticize manipulative willing and seek novel habits in its stead? If Hickman has it right, it should be possible to advance this through research, democracy, reshaping publics, and improved democratic participation. In any case, the dysfunctional characteristics of the manipulative ethos seem evident enough—even if it is not clear with what criteria we should assess functionality.

To put it in simple terms, it seems that humans in advanced societies should find a new way of willing, in addition to or instead of the manipulative willing. As we have seen, in its most comprehensive form the manipulative willing has the whole reality as its object. No criterion—or very few criteria—tell human beings what they are not allowed to manipulate or what direction the manipulation should take. The goals of manipulation—profit, health, pleasure, or novelties—of course set some boundaries for manipulation. But these boundaries are malleable and to an extent inherent in the manipulative willing.

In the course of this inquiry we encountered a few cues for a possible revision of the manipulative ethos. Hickman's evolutionary account underscored that human adaptation to habitats is both active manipulation of the environment and "passive" molding of human beings themselves in order to fit in the living conditions. The clue behind focal things and practices is that the otherness of things and temporal structure of practices shape human beings. Within focal practices, interacting with things, human beings give reality a say in their lives. Consciously and voluntarily accepting burdens, moreover, exhibits a novel way of willing. Organic life in Jonas's thinking has a parallel role: recognizing humans as organic beings beyond body-mind dualism will set technological activities on a humbler path. Jonas's recommendations in bioethics, for instance, point out what taking this organic reality into account should mean: limiting manipulation of and search for narrowly abstracted goods. Jonas reveals an organic being underneath the modern subject; Borgmann shows the primary thinghood of things underneath scientific objects and technological devices. All these strokes, however, address willing somewhat indirectly.

Let us consider a more historical perspective on willing. The modern turn in willing started from Machiavelli's new reading of politics.[40] Machiavelli rejected both theological and classical philosophical constraints for worldly willing. Instead of striving for "imaginary perfections" humans were to practice control over Fortune. Bacon and Descartes followed Machiavelli's move of freeing the political will and based their new science on the novel volitional stance: more than political fortune, nature was to be controlled through adequate scientific knowledge. If one wanted to repeat the earlier turn in an alternate or inverse way, one should find in the political realm an emerging new way of willing and use it as a lead for a turn in the will governing technoscience. I think if such a candidate exists, it might be "passive," or nonviolent, resistance used for political purposes. Consider Mohandas Gandhi's (1869–1948) work in the Indian struggle for independence, the civil rights movement in the United States, and the nonviolent revolutions in the Eastern Bloc countries in 1989. The form of willing in nonviolent resistance appears to be new. It is clearly different from the willing in earlier revolutions and civil wars of the modern times, of which the French Revolution with associated terror is the most emblematic example.[41]

What could we then learn from the passive resistance? To articulate it in a bold manner is not easy. Nonviolent resistance, nonetheless, exhibits a novel form of willing, and it has been successful. Instances of it have taken place earlier in history, but nonviolent resistance truly became widespread and victorious in the 20th century. Prior to Gandhi's efforts in South Africa and India, resignation and violent uprising seemed the only alternatives for the oppressed. Moreover, excluding the first Christian martyrs, Jesus' most radical teachings had not previously been put into practice. The successfulness of nonviolence makes it resistant to Machiavelli's critique of the imaginary ideals.[42] This success is proved by the dozens of bloodless revolutions that have taken place since the mid-20th century, roughly according to principles of nonviolence. Nonviolent resistance also discards technological means—weapons—in taking its stance, and its aim is not power in the usual sense, but rather not giving in to untruth and oppression—Gandhi's term *satyagraha* literally means "holding onto truth."[43]

Both Schopenhauer's metaphysics and the practice of political nonviolent resistance suggest a good beyond the values posited by the will. But how does this connect to technology? Technological action, making and using devices, is of course goal oriented and enabled by active willing. But the possibility of denying, or transcending, will shows that will to mastery and values embedded in it is not the only human reality, although reliance on will to power might still be the most "realistic" way to depict the world—as the modern thinkers from Machiavelli to Nietzsche have argued. These observations, nevertheless, show that the mastery-seeking will is not the only way to will.

It is not exactly clear what might be some concrete examples of this novel type of willing in the context of technology. Almost by definition "nonviolent resistance" needs an oppressor and a political conflict as its *raison d'être*. However, such movements as "voluntary simplicity" and "downshifting" might be domains where new ways of willing, akin to those within nonviolent resistance, are explored. "Simplicity" refers to reducing possessions, spending, and luxuries; "downshifting," to the same phenomena more from the viewpoint of pace and reducing time devoted to work. Reasons for these practices can today be philosophical, personal, spiritual, or ecological. Henry David Thoreau (1817–1862), the author of *Civil Disobedience* (1849) and *Walden* (1854), is one common father for both the nonviolent resistance and the simplicity movements.

What I have sketched with the help of Schopenhauer and nonviolence somewhat coheres with Heidegger's "releasement" (*Gelassenheit*) and Jacques Ellul's "ethics of non-power."[44] Moreover, my sketch also shares some of their overt generality on the topic. Therefore, taking the problem of willing seriously would mean more particular work at least on three different levels. First, the lead showed by thinkers who have elaborated on the role of the will could be followed more in philosophical reflection. This would mean more studies that set modern technoscience in a broad perspective in terms of will. Studying the metaphysical background of modern technoscience would be part of this endeavor. At the very least, backward-looking, historical, and deconstructive studies would be needed. Constructing new metaphysics that would amend the present situation, as we have seen, is a more precarious pursuit. Yet at some point this might be necessary. Second, forms of willing in technology should be studied carefully within the mid-level ethics: this means analyses of policies, institutional arrangements, communal practices, and education.[45] It seems clear, for instance, that current developments in both education and work life push the ethos of control, competition, and manipulation further. This is where Hickman's techno-logy could be at its most powerful: studying habitual patterns in volitions. Third, this observation raises a question about a way of life. As is known, Aristotle in *Nicomachean Ethics* argues that three ways of life are on offer for human beings: hedonistic life in which pleasure is the highest good, political life that seeks honor, and philosophical life, the highest good of which is wisdom or knowledge. He then adds a fourth way of life that revolves around money.[46] In these terms we are compelled to inquire into a way of life and a form of willing beyond the present technoscientific form of life. In a sense it would be a "post-technological" way of life.

Jonas's sketch of responsible life that does not seek for the *summum bonum* but, being aware of its own temporality, looks after fragile nature could well be read as a new way of life. Heuristics of fear and its more recent reading as the precautionary principle underscore the main feature of the new way of life: humans should humble down to face their own ignorance as well as the uncertainty and fragility inherent in the technoscientific systems. Borgmann's thinking on practices also points in the direction of a new way of life, although he does not push his argument far. Both Jonas and Borgmann seem to be taking their cues from ancient philosophical ideals and thinking through the appropriate revisions that they need in a technological age. Hickman does not address the content of a new way of life—perhaps he is not suggesting one. On the one hand, Hickman could well be read as advocating a new way of life that is based on the acknowledged absence of transcendent sources of value and all the related recastings. Hickman seems to think that far too many people, however, are somewhat unready to accept the pragmatist starting point and cling to old-fashioned sources of value. On the other hand, Hickman's account of technology can, more minimally, be read as a piecemeal method for correcting elements in a way of life: identifying habitual patterns, submitting them to critical reflection, and experimenting with them.

One last point in relation to the problem of will is the relevance of cultural dialogue. Cultural comparisons enable us to see Western culture more clearly in its distinctive features. Moreover, if the next step for us is less frenetic willing and transformation of the world, it is plausible to expect that other traditions have something to teach us. Comparing Borgmann's thinking with traditional Japanese culture confirmed this expectation in one case. Ordinary citizens in many ways seek answers to questions about ways of life from other traditions. Consider Asian philosophies and religions, yoga, meditation, martial arts, cuisines, dietetics, and alternative medicine. Addressing these phenomena may be a vexing task for a philosopher but should not be left untouched as pseudoscience. I primarily mean serious philosophical dialogue between East and West. Eastern philosophies and religions have cultivated practices of disciplining will, virtuous life, and self-knowledge for millennia.

NOTES

1. Thomas S. Kuhn, *The Copernican Revolution: Planetary Astronomy in the Development of Western Thought* (1985, first published 1957). Jacob Klein, "The Copernican Revolution," *Lectures and Essays* (1985), 85–115. Nicolaus Copernicus, *On the Revolutions of Heavenly Spheres* (1995, first published 1543), 3–30. Johannes Kepler, *Selections from Kepler's* Astronomia Nova (2008, original work 1609), 1–28.

2. René Descartes, *The Geometry of René Descartes* (1954, first published 1637), 1–37; *Discourse on Method* (2007, first published 1637 with *Geometry, Optics,* and *Meteorology*). Jacob Klein, *Greek Mathematical Thought and the Origin of Algebra* (1968), 197–211. Richard Kennington, *On Modern Origins: Essays in Early Modern Philosophy* (2004), 187–204.

3. In chapter XV of *The Prince* (1998) Machiavelli famously separated himself from what he saw as an idealist tradition of the ancients: "But since my intent is to write something useful to whoever understands it, it has appeared to me more fitting to go directly to the effectual truth of the thing than to the imagination of it. And many have imagined republics and principalities that have never been seen or known to exist in truth; for it is so far from how one lives to how one should live that he who lets go of what is done for what should be done learns his ruin rather than his preservation" (61). For a brief sketch of stages in modernity see Leo Strauss, "The Three Waves of Modernity," *Political Philosophy: Six Essays by Leo Strauss* (1975), 81–98.

4. Typical and preeminent examples of premodern thinking on reality as norm-giving for human beings include Stoicism, "Torah ontology" (thought that the created world embodies the commandments of Torah) of the wisdom books in the Hebrew Bible (see, e.g., Proverbs 8:22, 27, 32; Sirach 16:26–27), and Thomas Aquinas's (esp. *Summa Theologiae*, qq. 90–97) thinking on natural law.

5. This idea is mostly elaborated upon in Book I of *The New Organon.* "Human knowledge and power meet in one, for where the cause is not known the effect cannot be produced. Nature to be commanded must be obeyed, and that which in contemplation is as the cause is in operation as the rule" (*The New Organon*, I.3; see also I.129, II.1). "It is idle to expect any great advancement from the superinducing and engrafting of new things upon old. We must begin anew from the very foundations, unless we would revolve forever in a circle with mean and contemptible progress" (II.31).

6. Kristin Shrader-Frechette in *Risk and Rationality: Philosophical Foundations for Public Reforms* (1991, 38), for example, argues that "up to 90 percent of cancers in the United States are environmentally induced and hence theoretically preventable." See also *Risk and Rationality*, 27, 70; Shrader-Frechette, *Environmental Justice* (2002), 186–191, 200–202.

7. Max Weber, "Science as a Vocation," *From Max Weber: Essays in Sociology*, edited by H. H. Gerth and C. Wright Mills (1948), 143. "Was ist unter diesen inneren Voraussetzungen der Sinn der Wissenschaft als Beruf, da alle diese früheren Illusionen: "Weg zum wahren Sein", "Weg zur wahren Kunst", "Weg zur wahren Natur", "Weg zum wahren Gott", "Weg zum wahren Glück", versunken sind. Die einfachste Antwort hat Tolstoj gegeben mit den Worten: "Sie ist sinnlos, weil sie auf die allein für uns wichtige Frage: 'Was sollen wir tun? Wie sollen wir leben?' keine Antwort gibt." Die Tatsache, daß sie diese Antwort nicht gibt, ist schlechthin unbestreitbar. Die Frage ist nur, in welchem Sinne sie "keine" Antwort gibt, und ob sie statt dessen nicht doch vielleicht dem, der Frage richtig stellt, etwas leisten könnte" (Weber, *Gesamtausgabe.* Abteilung I, Band. 17, 1992, 93). Caleb Thompson's essay "Wittgenstein, Tolstoy and the Meaning of Life" (1997) treats insightfully the cited aspects of Tolstoy's thought and Wittgenstein in respects relevant to observations made here and in chapter 2.

8. "Science as a Vocation," esp. 139–141, 148, 155.

9. Rüdiger Safranski, *Martin Heidegger: Between Good and Evil* (1998), 145–275.

10. Existence, in fact, is capable of not only mistakes concerning knowledge and morals but existential falsehood as well. That is what Heidegger's fallenness, *Verfallenheit* (*Being and Time,* 1962, §§ 34b, 35–38, 68c), for instance, refers to.

11. Outside the frameworks discussed here, Emmanuel Levinas's approach (in *Totality and Infinity: An Essay on Exteriority*, 1969, 53–102, 194–220) to thinking about truth and discourse as based on the subject's more primary relationship to the Other is a strong argument that can find a way out of "Weber's problem." In brief, Levinas argues that the claim for justice that emerges from the face of the other precedes truth and is a necessary condition for it. Also, discourse with its requirement of objectivity presupposes the more fundamental relation to the other.

12. Langdon Winner, *Autonomous Technology: Technics-out-of-Control as a Theme in Political Thought* (1977), 229.

13. Aristotle, *Metaphysics*, 987a1–6.

14. See, e.g., *Apology*, 20d–e, 21d, 22d, 29b; *Crito*, 47d–48b; *Protagoras*, 313a–, 343b; *Phaedrus*, 229e; *Phaedo*, 96a–102b; Gregory Vlastos, *Socrates: Ironist and Moral Philosopher* (1991), 48, 53–56, 86–91.

15. See, e.g., Jacob Bronowski, *Science and Human Values* (1990).

16. On this development in bioethics: Margaret P. Battin, "Bioethics," *A Companion to Applied Ethics* (2003), 297–298, 307–311.

17. For a broader theoretical overview see Andrew Feenberg, *Questioning Technology* (1999), 101–129.

18. For a focused approach to this question see Topi Heikkerö, "Material Culture and the Conditions for the Good Life: Albert Borgmann on Thinking about Ethics in Technological Culture," *Praxiology and the Philosophy of Technology: The International Annual of Practical Philosophy and Methodology* (2008).

19. Larry A. Hickman, *Philosophical Tools for Technological Culture: Putting Pragmatism to Work* (2001), 133.

20. "Mid-level" here is used in a different sense than in Beauchamp and Childress's "Principlism" (*Principles of Biomedical Ethics,* 2001). In their sense "mid-level" refers to principles that are "above" casuistry but "below" general ethical theories (Donald C. Ainslie, "Principlism," *Encyclopedia of Bioethics*, 2003).

21. See, e.g., *Republic*, 522–541, 600b, 618b–619b; *Phaedo*, 64a–; *Phaedrus*, 249d– ; *Nicomachean Ethics*, I.5, X.6–9; *Politics*, VII.1–4. Pierre Hadot's reading of the ancient philosophical traditions in *What Is Ancient Philosophy?* (2002) emphasizes philosophy as a way of life.

22. See, e.g., Alasdair MacIntyre, *After Virtue: A Study in Moral Theory* (1984), 181–203; Michel Foucault, "Politics, Polemics and Problematizations: An Interview with Michel Foucault," *Ethics: Subjectivity and Truth (Essential Works of Michel Foucault, 1954–1984, Vol. 1)* (1997), 111–121.

23. In Finland 10–15 per 10,000 babies are born with Down syndrome; without screening and the possibility of the selective termination of pregnancy the number would be doubled. Thus, roughly half of babies with Down syndrome are aborted (statistics of National Institute of Health and Welfare, accessed February 21, 2009, www.stakes.fi/EN/tilastot/statisticsbytopic/reproduction/conganomaliessummary.htm). A rapid 50 percent decrease in Down birth within the Danish health care system is reported by Charlotte K. Ekelund, Finn Steren Jørgensen, Olav Jørn Petersen, Karin Sundberg, and Ann Tabor in "Impact of a New National Screening Policy for Down's Syndrome in Denmark: Population-Based Cohort Study" (*British Medical Journal*, 2008).

24. My description of the fetal screening in Finland is mostly based on Mianna Meskus's studies "Governing Risk through Informed Choice: Prenatal Testing in Welfarist Maternity Care" (*Contest Categories: Studies of the Life Sciences in Society*, 2009) and *Elämän tiede: Tutkimus lääketieteellisestä teknologiasta, vanhemmuudesta ja perimän hallinnasta* [Science of Life: A Study on Medical Technology, Parenthood, and the Management of Heredity] (2009), as well as Heli Pruuki's *Kuka on potilas? Suomalaisen sikiöseulonta käytännön tarkastelua* [Who Is the Patient? A Social-Ethical Study of the Finnish Practice of Prenatal Screening] (2007). See also Peter-Paul Veerbeek, "Obstetric Ultrasound and the Technological Mediation of Morality: A Postphenomenological Analysis," *Human Studies* (2008). Meskus's book looks at the longer history of selective abortions, sterilizations, and marriage prohibitions for disabled and mentally ill people as the background of the present fetal screening processes. She (169–185, esp. 173) sees the development of these practices in the 20th century as a process of purification (see Bruno Latour, *We Have Never Been Modern*, 1993). In the process, health care policies that before WWII aimed at "the strength of the nation" and, later, at cost efficiency in public health care have been purified into "purely technical" means that can provide autonomy for the parents. Following Latour, Meskus shows how such purification fails: the technical and the political remain entangled. Meskus (137) also points out a case of reverse adaptation. In the 1990s new tests for finding markers indicating the fetus's disability in pregnant mother's blood were included in the screening processes. According to her reading of the research literature, this happened solely because the technical possibility existed. The public health officials' and physicians' eagerness in preventing disabilities had already decreased to a great extent.

25. In the Nordic countries the situation is varied in the following manner: In Finland all pregnant mothers are screened, unless they decide not to participate in the screenings. In Sweden all the mothers have the right to tests if they themselves ask for them. Norwegian policy assumes that individuals with Down syndrome can live good enough lives that there is no need to prevent this form of disability (Meskus, *Elämän tiede*, 120–123).

26. Pruuki, *Kuka on potilas?*, esp. 9–10, 184–194. Meskus, *Elämän tiede*, 103–138.

27. Both Meskus's and Pruuki's studies reveal the paradoxical character of the private and autonomous choice. If the parents are "left alone" to decide the fate of their fetus—as the official policy requires—they face a dilemma. When making a purely private choice they have few means to assess whether the choice is right. But if they seek such criteria the private and autonomous character of their choice is in danger. In general, Meskus and Pruuki suggest that parents would need more help in tackling the anxiety of the choice. This help obviously should come from the mid-level. Another problem, of course, is the fact that parents are usually not left alone but receive implicit and explicit messages about the correct, usual, or normal choice in their situation. Making the choice to knowingly have a disabled baby private at least indirectly suggests to the parents that they alone will carry the responsibility of having a child that needs extra help.

28. Matthew B. Crawford in *Shop Class as Soulcraft: An Inquiry into the Value of Work* (2009, 126–160) gives an elaborate account of the recent developments.

29. Committee on Science, Engineering, and Public Policy, *On Being a Scientist: Responsible Conduct in Research* (1995). National Advisory Board for Research Ethics, *Good Scientific Practice and Procedures for Handling Misconduct and Fraud in Science* (2002).

30. This account is based on my experiences in teaching research ethics and bioethics classes in Finland.

31. In addition to the social movements mentioned, there have been more specifically technology-related movements that have influenced respective developments. Consider consumer protection movements, the antinuclear (both weapons and power) movement, criticisms of biomedical research and therapies, and the concerns expressed by computer professionals. Their effect, however, has not been of the same transforming magnitude as that of the feminism or labor movement.

32. For an overview, see Carl Mitcham, *Thinking through Technology: The Path between Engineering and Philosophy* (1994), 247–266.

33. Francis Bacon, *The New Organon*, I.129 in *Selected Philosophical Works* (1999), 147.

34. René Descartes, *Passions of the Soul*, 50, 147–161 in *The Philosophical Writings of Descartes*, Vol. I. (1985), 348, 381–388 (AT, 368–370, 440–454).

35. *The World as Will and Representation*, Vol. 1, §§ 21–27. In very brief, Schopenhauer argues that will (*Wille*) is in fact Kant's thing in itself (*Ding an sich*) and can be known by self-perception of one's own active willing. Consequently, it is also possible to observe will in nature. Will in Schopenhauer's thinking is one, undifferentiated. Differentiation comes from forms of perception, understood roughly in Kantian manner.

36. *The World as Will and Representation*, Vol. 1, §§ 64–68, 70–71; *On the Basis of Morality*, §§ 12–19, 22.

37. Hans Jonas, "Biological Foundations of Individuality," *Philosophical Essays: From Ancient Creed to Technological Man* (1974), 185–205.

38. Mitcham, *Thinking through Technology*, 259. Above the quoted passage, Mitcham points out shortcomings of the traditional ethics in addressing volitional stance underlying modern technology: "First, traditional moral philosophy fails to provide an adequate account of those technological volitions described by phrases such as 'will to control' or 'will to power' or even 'pursuit of efficiency.' Second, although traditional ethical analyses consider the relationship between different understandings of the good and certain human institutions (states, educational curricula, family structures, economic systems, etc.), they do not, except in quite limited ways begin to address the correspondence between different understandings of the good and technology" (259).

39. Martin Heidegger, *Discourse on Thinking: A Translation of* Gelassenheit (1966); *The Question Concerning Technology and Other Essays* (1977). C. S. Lewis, *The Abolition of Man* (1943). G. H. von Wright, *Vetenskapen och förnuftet. Ett försök till orientering* [Science and

Reason: An Attempt at Orientation] (1986). Pierre Hadot in *The Veil of Isis: An Essay on the History of the Idea of Nature* (2006, 91–151) calls the manipulative ethos toward nature the Promethean Attitude. Its opposite and counterpart is the Orphic Attitude (91–98, 155–229), which approaches nature with awe and respects nature's mysteries. For Hadot, Plato's *Timaeus* is an emblematic example (155–159, 182–183) of the Orphic Attitude, as are Goethe's studies on nature (148–150). Hadot's account, however, depicts both attitudes as having begun in the Greek thinking.

40. Of course, "starting" is always relative. For example, voluntaristic theology and metaphysics of Johannes Duns Scotus (1265–1308) and William of Ockham (1288–1348) could be seen as a crucial earlier turning.

41. The roots of the ideas on which nonviolent resistance is based are in the New Testament, primarily in Jesus' Sermon on the Mount (Matthew 5–7, esp. 5:38–48). In the modern context the tradition, however, in a way stems from Schopenhauer, whose thought was picked up by Tolstoy. Tolstoy's rereading of the New Testament had a crucial impact on Mohandas Gandhi (*An Autobiography: The Story of My Experiments with Truth*, 1957, 137, 160). It thus seems that philosophical background for passive resistance, at least to an extent, is Schopenhauer's doctrine of the denial of the will and related compassion.

42. I do not, however, think that Machiavelli's "pragmatic" critique of ancient Greek philosophy, primarily Plato and Aristotle, is damning for "idealistic" rationalism. Another way (in addition to searching for something new) to think about amending the modern predicament is to try to revive neglected parts of the ancient philosophical traditions.

43. "Its root meaning is holding on to truth, hence truth-force. I have also called it Love-force or Soul-force. In the application of Satyagraha I discovered in the earliest stages that pursuit of truth did not admit violence being inflicted on one's opponent but that he must be weaned from error by patience and sympathy. For what appears to be truth to the one may appear to be error to the other. And patience means self-suffering. So the doctrine came to mean vindication of truth not by infliction of suffering on one's opponent but on one's self" (Mohandas Gandhi, *Non-Violent Resistance (Satyagraha)*, 2001, 6). A Schopenhauerian reading of Gandhi's statement could go along the following lines: In *satyagraha* actors to an extent bracket their own values and do not impose them on their opponents. Taking the suffering on themselves, they make the opponent see beyond valuations, and more fundamentally beyond the rigid principle of individuation. The truth and good that are realized in this process are something that is beyond the values that are functions of willing.

44. Jacques Ellul, "The Power of Technique and the Ethics of Non-Power," *Myths of Information: Technology and the Postindustrial Culture* (1980); "The Ethics of Non-Power," *Ethics in an Age of Pervasive Technology* (1980).

45. For a humble effort to address willing in engineering education, see Topi Heikkerö, "Ethos in Engineering Education: How to Address the Volition Dimension of the Engineer's Social Responsibility?" *European Journal of Engineering Education* (2008).

46. *Nicomachean Ethics*, I.5.

Chapter Seven

Conclusions

This inquiry has examined the question: *How to think about ethics in a technological world?* I divided this question into a thematic part that framed the problem of ethical thinking in technologized conditions and a descriptive part that read three central approaches for addressing the problem. I began by examining the relationship of ethics and technology in the contemporary world. This relationship appears to have a threefold form. I first looked at the ways in which technological advance poses challenges for ethical thinking. Some of these challenges can be addressed effectively by specialized ethical disciplines such as bioethics and engineering ethics. Some of the challenges are broader and politically more complicated. Just distribution of the costs, harms, benefits, and risks of technological projects is a central example of these broader questions. A second perspective concerned the applicability of traditional ethical concepts in technologically altered conditions. Ethics in many respects derives its fundamental notions from the centuries before technological modernity. Most crucially, I questioned the assumption that neither technological production nor scientific knowledge acquisition is really morally laden. A third viewpoint concerned the possibility—or better, difficulty—of finding a foundation or grounding for ethics. In brief, the problem of grounding revolves around the character of technoscientific rationality. The huge advance in science and technology has resulted in technoscientific rationality becoming the measure of reason. Ethics as a normative discourse on human affairs appears unscientific, unmeaningful, and without a proper foundation when measured by scientific criteria. This issue has been addressed by both analytic and phenomenological tradition in philosophy, but in very different ways. In conclusion I argued that philosophy of technology needs to take an interdisciplinary approach in order to address the ethical problems reviewed. Philosophers of technology should make use of the work done in applied ethics as well as in science and technology studies. Thus far these three research programs have not engaged in much dialogue, which is understandable, as their basic starting points clearly differ. I argued for the soundness of the underlying assumptions in philosophy of technology and suggested that the interdisciplinary project mainly means philosophy of technology being supplemented by knowledge acquired through the other two research programs. Moreover, I argued for understanding philosophy in broad and traditional terms—in contrast to understanding philosophy of technology as a novel, specialized field. Philosophy in its traditional sense inquires into the first principles of thinking, comprehensive ways of life, and how the good is understood within them.

The descriptive part began by taking Hans Jonas's thinking on technology under scrutiny. I first described Jonas's analysis of the modern worldview. His analysis revolved around nihilism: in the modern condition human beings find themselves in a meaningless, alien world. Value and meaning have become solely dependent on human attribution. At the root of this nihilistic situation, Jonas found what he perceived as the inherent dualistic structure of the Western metaphysics: the human conscious being is always cut off from the material world. A central motive in Jonas's reconstructive thinking on the ontology of organic life was to overcome both dualism and nihilism. This he aimed to do by showing that on the level of organism the distinction of consciousness and matter cannot be made, and that being and good inseparably coexist on that level as well. This ontological grounding of ethics served as a starting point for Jonas's ethics of technology, which he presented under the rubric of the Imperative of Responsibility: "Act so that the effects of your actions are compatible with the permanence of genuine human life." Both a strength and central problem of Jonas's approach was his reliance on metaphysical thinking: in addition to being able to touch upon worldviews in a comprehensive manner, it made Jonas's approach difficult to assess critically.

The second thinker under scrutiny was Albert Borgmann. First, I looked at his analysis of the technological way of taking up with reality, the device paradigm. The device paradigm aims at disburdening human life from menial labor, toil, danger, and sickness. It has succeeded. The downside of the device paradigm, however, is its tendency to structure life so that dichotomies such as means-ends, body-mind, individual-community, and work-leisure sharpen. This sharpening results in disburdened life becoming shallow, disintegrated, and not truly meaningful. To amend this malaise Borgmann introduced focal things and practices. They are mostly old-fashioned things—as contrasted to commodities and objects—and demanding activities. Focal things and practices cannot be managed within the technological pattern. Borgmann made use of these concepts in his broader social analysis and argumentation.

Third, I focused on Larry Hickman's pragmatist philosophy of technology. In Hickman's thinking, the distinction of human action into habitual and reflective was central. Habitual action makes use of techniques, patterns of action, unreflectively; techniques serve as platforms in action. Habitual action is necessary in life; human life, in fact, is mostly habitual. Habitual action, however, turns dysfunctional if the conditions become unfavorable for it. Then reflective action needs to take place. Technological, or reflective, action in Hickman's thinking is action that through experimentation and reflection aims at improving techniques and habits. All conscious improvement of unreflected patterns of action is technology for Hickman. Thus, in addition to research, development, and industrial production, art and improving democratic procedures in decision-making are technology. Hickman, moreover, wants to recast the distinctions between *praxis*, *poiesis*, and *theoria.* According to him, all human action has moral, productive, and epistemic elements. From this it follows that ethical inquiry does not exist in a special realm but is in the same continuum with all technological action. At the same time, no human action, according to Hickman, is morally neutral.

In the previous chapter I brought together questions from the earlier chapters. My argument was that the broad frame for ethics of technology is the rethinking of the modern project. I first looked at the problematic separation of being and good through comparing how the three approaches treat the concept of value and how they tackle the problem of modernity as described by Max Weber. Then I clarified what I see as the best way to think about *praxis*, *poiesis*, and *theoria* in contemporary conditions. The main point was that these three categories are better kept separate than blurred, but they are in need of serious revision, as the current forms of *poiesis* and *theoria* have little grounds to claim value neutrality. Next I presented arguments in order to underscore the relevance of the mid-level for the ethics of technology. Drawing on earlier chapters I claimed that many crucial questions are settled at the mid-level—without much discussion or full awareness—rather than by the choices of separate individuals or on the macro-level that asks questions about the principles underlying a just society. I went on to describe what I see as the minimum task of the ethics of technology. The minimum task does not challenge the basic practices of the contemporary advanced societies but aims to practice clear ethical thinking within them. Hence it is for the most part *procedural* in character. In contrast to minimum task, I presented a few strokes in the direction of a maximal ideal for ethics of technology. In its more serious form philosophy of technology should inquire into willing, or volition, underlying the modern technological project. This direction of inquiry was outlined by Heidegger in the mid-20th century but has not yet been picked up by the philosophy of technology community. I argued that inquiry into the form of willing in technology equals the traditional philosophical inquiry into the first principles, as in the modern technoscientific condition the intellectual principles have become subservient to willing.

The central conclusions of this inquiry can be summarized under five rubrics. The first is the above plea for an interdisciplinary character of philosophy of technology and the need to conceive philosophy in a comprehensive sense. Second, in chapter two I questioned the feasibility of the traditional distinction of human action into *praxis*, *poiesis*, and *theoria.* This distinction was, furthermore, strongly criticized by Borgmann and Hickman, and I found their criticisms valid and potent. The crux of this claim is that, at least in contemporary conditions, moral deliberation, material production, and scientific description of reality cannot be kept separate. Most importantly, technological artifacts structure the world and the human being-in-the-world in such ways that it is impossible to consider them morally neutral means. Moreover, as Hickman points out, theoretical concepts work as tools in both production and moral thinking. Therefore, knowledge creation has productive and moral elements in it: it creates conceptual tools and its results affect human orientations. My own conclusion, however, was that these categories are worth keeping separate with adequate revisions and that *praxis* and *theoria* are in need of strengthening.

Tightly connected to the above is the third point: this study continually encountered the problematic character of the fact-value distinction and awkwardness of the concept of value itself. Borgmann managed to skillfully elaborate the structure of value talk behind the simpleminded expectation of moral elevation. He showed that the space for morality is, in fact, rather small in value talk, as the pattern of technology settles many questions even before they are articulated. Jonas's analysis emphasized the arbitrariness of values if they are solely dependent on valuation acts of human consciousness. As already mentioned, Hickman's thinking on making, doing, and knowing is insightful, although his line of thinking comes with some costs. Hickman's relational pragmatist ethics appears thin and one-dimensional: if ethics is fully flexible and open for all the possible experiential revisions, it is conceptually incomplete. Either it is empty of final criteria or these criteria are left unarticulated.

As a separate, fourth point, it behooves us to note that the problem of nihilism remains unresolved. The problem was framed and analyzed from a few viewpoints in the context of technology. Briefly, the problem is that if values depend on human choice they have little or no binding force. If something is good only because one chooses to value it, it does not have much commanding authority, since it is by definition open for reconsideration, a new value choice. No philosopher's stone, real source of value, or way of thinking and speaking about morals, however, is obvious. Exotic but promising possibilities for thinking were opened by Jonas, who questioned the modern dogma that being in itself is devoid of good—that being simply is, but good needs to be "added" to being through laborious argumentation. His constructions, which include good in being, I found less convincing than his critical analyses. Borgmann stressed the role of things in orientation in the world. Things are present for humans in a commanding manner. They are able to resist human pursuits of control and manipulation; they remain other to human beings. In their ability to focus and orient life, things are somewhat analogous to values. But things are material, and not fully graspable within the conceptual discourse—that is why they are things; otherwise, they would be objects. Values, to the contrary—at least in the common modern way of understanding them—are ideal and therefore manageable with concepts. While values are created by valuations, things limit choices. In chapter six I suggested that the most appropriate way to approach the problem of nihilism is careful study of willing in technoscientific action and the self-knowledge arrived at in this manner. Both Schopenhauer's ethics based on his metaphysics and the practice of nonviolent resistance suggest that ethically guiding phenomena—primarily compassion (*Mitleid*)—occur beyond the value-positing willing.

The fifth and final point is the request to study ethics on the meso-level. The meso-level is the domain of moral questions between microethics, which focuses on particular, instantaneous, and individual moral choices, and macroethics, which discusses general moral principles in societies. Of the micro viewpoint, some applied ethics cases might be good examples: If the hospital has only one respirator and three patients need it, to whom should it be given? Of the macro perspective, John Rawls's theory of justice could be a good illustration. In health care, the macro perspective would concern the just distribution of health care resources within a society. Meso-level ethics looks at longer time frames than microethics and is interested in policies, institutions, habits, and practices. Meso-level would, in the case of health care, be interested in practices within health care institutions. Borgmann looked at the world through the lens of meso-level ethics both in his analysis of the silent structuring power of the device paradigm and in his idea of focal things and practices as the vehicles for a reform in contemporary societies. In Hickman's philosophy of technology, meso-level ethics is always implicitly present, since his techno-logy is conscious, experiential reflection on techniques, habits, and institutions—meso-level platforms for human action.

This study thus puts forth arguments in favor of taking the meso-level into account, alongside the micro- and macroethical considerations. These ideas obviously are not entirely new. They have been present in the philosophical tradition for most of its history—both in the West and in the East. In recent discussions, thoughts along these lines have been developed by Alasdair MacIntyre and Michel Foucault (1926–1984). MacIntyre's work has been central in the revival of the virtue ethics. Foucault analyzed the formation of subjectivity in his genealogical studies and hinted at reformatory directions in his writings on the technologies of the self (*techniques du soi, practiques du soi*).[1] The present study underlines the relevance of the meso-level for the ethics of technology.

The need to sufficiently notice the meso-level, moreover, shows clearly why an interdisciplinary approach is necessary in addressing ethics in technology—the first point above. Since applied ethics often does not pay attention to practices and material structures as the broader moral contexts, it needs to be supplemented by a more sociological approach to technology. Since science and technology studies emphasize the neutral description of techno-social phenomena, they need to be supplemented by applied ethics that emphasizes human agency and a philosophy of technology that consists of more comprehensive and critical inquiry.

Let me end with one final note concerning future studies in ethics of technology. In recent decades Heidegger's and Ellul's thinking has often been termed "classical philosophy of technology." This term mainly refers to the comprehensive and unfalsifiable character of their views. Denying the need for empirical knowledge and practical engagement in ethics of technology would clearly be foolish. Archiving Heidegger and others as the greats of the past and focusing on doing real, academic research, however, brushes many of the original philosophical questions about technology aside. Here I brought up will as a central question. "Classical," in fact, might not be a pejorative term at all: reclaiming philosophy in a more classical sense seems necessary for philosophy of technology, if it is to be a transforming agent in our trying times. By "more classical" I mean philosophy that is rooted in the search for self-knowledge, inquires into the first principles of thinking and science, and is able to envision and weigh comprehensive ways of life.

NOTE

1. MacIntyre, *After Virtue: A Study in Moral Theory* (1984); "Does Applied Ethics Rest on a Mistake?" *Monist* (1984). Foucault, "Technologies of the Self," *Ethics: Subjectivity and Truth (Essential Works of Michel Foucault, 1954–1984, Vol. 1)* (1997), 223–251.

Bibliography

Accident Investigation Board Finland. Investigation Report, A 1/2004 Y. Abridged English translation of the original Finnish report, 2004.

Achterhuis, Hans, ed. *American Philosophy of Technology: The Empirical Turn.* Bloomington, IN: Indiana University Press, 2001.

Adams, Henry. *The Education of Henry Adams.* New York: Modern Library, 1946.

Aeschylus. *Prometheus Bound.* Translated by David Grene. In *Aeschylus II*, edited by David Grene and Richmond Lattimore. 2nd Edition. Chicago: University of Chicago Press, 1991.

Aho, Tuomo, and Mikko Yrjönsuuri, eds. *Norms and Modes of Thinking in Descartes.* Helsinki: Acta Philosophica Fennica, 1999.

Ainslie, Donald C. "Principlism." In *Encyclopedia of Bioethics*. 3rd Edition. New York: Macmillan, 2003.

Airaksinen, Timo, and Wojciech W. Gasparski, eds. *Praxiology and the Philosophy of Technology: The International Annual of Practical Philosophy and Methodology*, Vol. 15. New Brunswick and London: Transaction Publishers, 2008.

Almond, Brenda, ed. *Introducing Applied Ethics.* Oxford and Cambridge, MA: Blackwell, 1995.

Almond, Brenda. "Introduction: Ethical Theory and Ethical Practice." In *Introducing Applied Ethics*, edited by Brenda Almond. Oxford and Cambridge, MA: Blackwell, 1995.

Anscombe, G. E. M. *Ethics, Religion and Politics: The Collected Philosophical Papers*, Vol. III. Oxford: Blackwell, 1981.

Apel, Karl-Otto. "The Problem of a Macroethic of Responsibility to the Future in the Crisis of Technological Civilization: An Attempt to Come to Terms with Hans Jonas's 'Principle of Responsibility.'" Translated by Wilson Brown. *Man and World* 20 (1987), 3–40.

Aquinas, Thomas. *Basic Writings of Saint Thomas Aquinas*, edited by Anton C. Pegis. Indianapolis: Hackett, 1997.

Arendt, Hannah. *The Human Condition.* Chicago and London: University of Chicago Press, 1958.

———. *Eichmann in Jerusalem: A Report on the Banality of Evil.* Revised and Enlarged Edition. New York: Penguin, 1994.

Aristotle. *Nicomachean Ethics.* Translated by H. Rackham. Loeb Classical Library. Cambridge, MA: Harvard University Press, 1947.

———. *Metaphysics*, Books 1–9. Translated by Hugh Tredennick. Loeb Classical Library. London: Heinemann, 1968.

———. *Metaphysics*, Books 10–14. Translated by Hugh Tredennick. *Oeconomica & Magna Moralia.* Translated by G. Cyril Armstrong. Loeb Classical Library. Cambridge, MA: Harvard University Press, 1969.

———. *On the Soul, Parva Naturalia, and On Breath.* Translated by W. S. Hett. Loeb Classical Library. Cambridge, MA: Harvard University Press, 1975.

———. *Metaphysics*. Translated by Joe Sachs. Santa Fe: Green Lion Press, 1999.

———. *Nicomachean Ethics.* Translated by Joe Sachs. Newburyport, MA: Focus, 2002.

Armstrong, Susan J., and Richard G. Botzler, eds. *Environmental Ethics: Divergence and Convergence.* New York: McGraw Hill, 1993.

Arnhart, Larry. "Human Nature." In *Encyclopedia of Science, Technology, and Ethics*. Detroit: Macmillan Reference, 2005.

Attfield, Robin. *Environmental Ethics: An Overview for the Twenty-First Century.* Cambridge: Polity Press, 2003.

Ayer, Alfred J. *Language, Truth and Logic.* London: Victor Gollancz Ltd., 1936.

Bacon, Francis. *The New Organon and Related Writings*, edited by Fulton H. Anderson. New York: The Liberal Arts Press, 1960.

———. *Selected Philosophical Works*, edited by Rose-Mary Sargent. Indianapolis and Cambridge: Hackett, 1999.

Baldine, Joanne. "Larry Hickman and Tuning Up the Technological Culture." *Techné* 7 (2003, No. 1), 12–25.

Battin, Margaret P. "Bioethics." In *A Companion to Applied Ethics*, edited by R. G. Frey and Christopher Wellman, 295–312. Oxford: Blackwell, 2003.

Bauer, Susanne, Hanne Jenssen, and Ayo Wahlberg, eds. *Contest Categories: Studies of the Life Sciences in Society.* Aldershot: Ashgate, 1999.

Beauchamp, Tom L. "The Nature of Applied Ethics." In *A Companion to Applied Ethics*, edited by R. G. Frey and Christopher Wellman, 1–16. Oxford: Blackwell, 2003.

Beauchamp, Tom L., and James F. Childress. *Principles of Biomedical Ethics*. 5th Edition. Oxford and New York: Oxford University Press, 2001.

Beauchamp, Tom L., and Ruth R. Faden. "Informed Consent." In *Encyclopedia of Bioethics*. 3rd Edition. New York: Macmillan, 1995.

Bernstein, Richard J. "Rethinking Responsibility." In *Hastings Center Report* 27 (1995), 13–20.

———. *Radical Evil: A Philosophical Interrogation.* Cambridge: Polity Press, 2002.

Bijker, Wiebe. "Do Not Despair: There Is Life after Constructivism." *Science, Technology, and Human Values* 18 (1993, No. 1), 113–138.

———. *Of Bicycles, Bakelites, and Bulbs: Toward a Theory of Socio-Technical Change.* Cambridge, MA, and London: MIT Press, 1995.

———. "Understanding Technological Culture through a Constructivist View of Science, Technology, and Society." In *Visions of STS: Counterpoints in Science, Technology, and Society Studies*, edited by Stephen H. Cutcliffe and Carl Mitcham, 19–34. Albany: State University of New York Press, 2001.

Bland, Kalman P. "Reflections on the Place of Gnosticism and Ethics in the Thought of Hans Jonas." In *The Legacy of Hans Jonas: Judaism and the Phenomenon of Life*, edited by Hava Tirosh-Samuelson and Christian Wiese, 283–291. Leiden and Boston: Brill, 2008.

Bloom, Allan. *Closing of the American Mind.* New York: Simon & Schuster, 1987.

Borgmann, Albert. *The Philosophy of Language: Historical Foundations and Contemporary Issues.* The Hague: Martinus Nijhoff, 1974.

———. *Technology and the Character of Contemporary Life: A Philosophical Inquiry.* Chicago: University of Chicago Press, 1984.

———. *Crossing the Postmodern Divide.* Chicago: University of Chicago Press, 1992.

———. "Reply to Larry Hickman." *Research in Philosophy and Technology* 12 (1992), 345–347.

———. "The Moral Assessment of Technology." In *Democracy in Technological Society: Philosophy and Technology*, Vol. 9, edited by Langdon Winner, 207–213. Dordrecht: Kluwer, 1992.

———. "Finding Philosophy." In *Falling in Love with Wisdom: American Philosophers Talk about Their Calling*, edited by David D. Karnos and Robert G. Shoemaker, 157–160. New York and Oxford: Oxford University Press, 1993.

———. "The Moral Significance of the Material Culture." In *Technology and the Politics of Knowledge*, edited by Andrew Feenberg and Alistair Hannay, 85–97. Bloomington and Indianapolis: Indiana University Press, 1995.
———. *Holding On to Reality: The Nature of Information at the Turn of the Millennium.* Chicago: University of Chicago Press, 1999.
———."A Scarcity of Focal Things: Reply to Pieter Tijmes." *Technology in Society* 21 (1999), 191–199.
———. "Heidegger and Ethics beyond the Call of Duty." In *Appropriating Heidegger*, edited by James E. Faulconer and Mark A. Wrathall, 68–82. Cambridge: Cambridge University Press, 2000.
———. "Reply to My Critics." In *Technology and Good Life?*, edited by Eric Higgs, Andrew Light, and David Strong, 341–370. Chicago: University of Chicago Press, 2000.
———. "Society in the Postmodern Era." *The Washington Quarterly* 23 (2000, No. 1), 189–200.
———. "Response to My Readers." *Techné* 6 (2002, No. 1), 110–125.
———. "Kinds of Pragmatism." *Techné* 7 (2003, No. 1), 26–35.
———. *Power Failure: Christianity in the Culture of Technology.* Grand Rapids: Brazos Press, 2003.
———. "Material Culture." In *Encyclopedia of Science, Technology, and Ethics*. Detroit: Macmillan Reference, 2005.
———. *Real American Ethics: Taking Responsibility for Our Country.* Chicago and London: University of Chicago Press, 2006.
Bostrom, Ann. "Risk Perception." In *Encyclopedia of Science, Technology, and Ethics*. Detroit: Macmillan Reference, 2005.
Brandt, Richard B. "Emotive Theory of Ethics." In *Encyclopedia of Philosophy.* New York and London: Collier Macmillan Publishers, 1967.
Brante, Thomas, Steve Fuller, and William Lynch, eds. *Controversial Science: From Content to Contention.* Albany: State University of New York Press, 1993.
Brittan, Gordon. "Technology and Nostalgia." In *Technology and Good Life?*, edited by Eric Higgs, Andrew Light, and David Strong, 70–88. Chicago: University of Chicago Press, 2000.
Bronowski, Jacob. *Science and Human Values*. Revised Edition. New York: Harper Perennial, 1990.
Buber, Martin. *I and Thou.* Translated by Walter Kaufman. New York: Touchstone, 1996.
Bugbee, Henry Greenwood. *The Inward Morning: A Philosophical Exploration in Journal Form*. Athens, GA: Georgia University Press, 1999.
Byerly, Radford Jr. "Space Shuttles *Challenger* and *Columbia* Accidents." In *Encyclopedia of Science, Technology, and Ethics*. Detroit: Macmillan Reference, 2005.
Callahan, Daniel. "Bioethics." In *Encyclopedia of Bioethics*. 4th Edition. New York: Macmillan, 2003.
Carson, Rachel. *Silent Spring*. 25th Anniversary Edition. Boston: Houghton Mifflin Company, 1987.
Cayley, David. *Ivan Illich in Conversation.* Concord: House of Anansi Press, 1992.
Cérézuelle, Daniel. "The Experience of Beauty in Nature and Its Anthropological Importance." *Colorado School of Mines Quarterly* 100 (2000, No. 2), 9–16.
Chadwick, Ruth. "Applied Ethics." In *Encyclopedia of Science, Technology, and Ethics*. Detroit: Macmillan Reference, 2005.
Collins, Harry, and Trevor Pinch. *The Golem: What You Should Know about Science*. 2nd Edition. Cambridge: Cambridge University Press, 1998.
———. *The Golem at Large: What You Should Know about Technology.* Cambridge: Cambridge University Press, 1998.
———. *Dr. Golem: How to Think about Medicine.* Chicago: University of Chicago Press, 2005.
Committee on Science, Engineering, and Public Policy. *On Being a Scientist: Responsible Conduct in Research.* Washington, DC: National Academy Press, 1995.

Copernicus, Nicolaus. *On the Revolutions of Heavenly Spheres.* Translated by Charles Glenn Wallis. Amherst, NY: Prometheus Books, 1995.
Copleston, Frederick. *A History of Philosophy, Vol. 7: From the Post-Kantian Idealists to Marx, Kierkegaard, and Nietzsche.* New York: Doubleday, 1965.
Corrado, Michael, L. *The Analytic Tradition in Philosophy.* Chicago: American Library Association, 1975.
Crawford, Matthew B. *Shop Class as Soulcraft: An Inquiry into the Value of Work.* New York: Penguin, 2009.
Crigger, Bette-Jane, ed. *Cases in Bioethics: Selections from the Hastings Center Report.* 3rd Edition. Boston and New York: Bedford/St. Martin's, 1998.
Cutcliffe, Stephen H. "Science, Technology, and Society Studies." In *Encyclopedia of Science, Technology, and Ethics*. Detroit: Macmillan Reference, 2005.
Cutcliffe, Stephen H., and Carl Mitcham, eds. *Visions of STS: Counterpoints in Science, Technology, and Society Studies.* Albany: State University of New York Press, 2001.
Darwall, Stephen. "Ethical Theory." In *Encyclopedia of Philosophy: Supplement.* New York: Macmillan Reference, 1996.
———. "Theories of Ethics." In *A Companion to Applied Ethics*, edited by R. G. Frey and Christopher Wellman, 17–36. Oxford: Blackwell, 2003.
Davis, Michael. *Thinking Like an Engineer: Studies in the Ethics of Profession.* Oxford and New York: Oxford University Press, 1998.
De George, Richard T. *Business Ethics.* 5th Edition. Upper Saddle River, NJ: Prentice Hall, 1999.
DeMarco, Joseph P., and Richard M. Fox, eds. *New Directions in Ethics: The Challenge of Applied Ethics.* New York and London: Routledge and Kegan Paul, 1986.
Descartes, René. *The Geometry of René Descartes.* Translated by David Eugene Smith and Marcia L. Latham. New York: Dover, 1954.
———. *The Philosophical Writings of Descartes*, Vol. I. Translated by John Cottingham, Robert Stoothoff, and Dugald Murdoch. Cambridge: Cambridge University Press, 1985.
———. *Discours de la méthode.* Paris: GF Flammarion, 1988.
———. *Œuvres philosophiques: Tome I (1618–1637)*, edited by Ferdinand Alquie. Paris: Éditions Garnier, 1988.
———. *Discourse on Method*, edited by Pamela Kraus and Frank Hunt. Translated by Richard Kennington. Newburyport, MA: Focus, 2007.
Dewey, John. *Studies in Experimental Logic.* Chicago: University of Chicago Press, 1916.
———. *The Quest for Certainty: A Study of the Relation of Knowledge and Action.* New York: Milton, Balch & Company, 1929.
———. *Logic: The Theory of Inquiry.* London: Allen & Unwin, 1939.
Didier, Christelle, and Bertrand Hériard Dubreuil. "Engineering Ethics: Europe." In *Encyclopedia of Science, Technology, and Ethics*. Detroit: Macmillan Reference, 2005.
Dogen, Eihei. *Moon in a Dewdrop: Writings of Zen Master Dogen.* Translated by Kazuaki Tanahashi. Canada: HarperCollins Canada, 1985.
Dostoevsky, Fyodor. *Crime and Punishment.* Translated by Richard Pevear and Larissa Volokhonsky. New York: Everyman, 1993.
———. *Notes from Underground.* Translated by Richard Pevear and Larissa Volokhonsky. New York: Vintage Books, 1993.
———. *Demons.* Translated by Richard Pevear and Larissa Volokhonsky. New York: Vintage Books, 1995.
Downey, Gary, and Juan Lucena. "Engineering Cultures." In *Science, Technology, and Society: An Encyclopedia*, edited by Sal Restivo, 124–130. Oxford and New York: Oxford University Press, 2005.
Dreyfus, Hubert L. "Heidegger on the Connection between Nihilism, Art, Technology, and Politics." In *The Cambridge Companion to Heidegger*, edited by Charles B. Guignon, 289–316. Cambridge: Cambridge University Press, 1993.
Dreyfus, Hubert L., and Charles Spinosa. "Further Reflections on Heidegger, Technology, and the Everyday." *Bulletin of Science, Technology, and Society* 23 (2003, No. 5), 339–349.

Durbin, Paul T. "Philosophy of Technology." In *Technology and Good Life?*, edited by Eric Higgs, Andrew Light, and David Strong, 38–49. Chicago: University of Chicago Press, 2000.

———. "Introduction." *Techné* 7 (2003, No. 1), 1–11.

———. "Philosophical Tools or Technological Culture: Comments from an Activist Perspective." *Techné* 7 (2003, No. 1), 36–41.

———. "Pragmatism." In *Encyclopedia of Science, Technology, and Ethics*. Detroit: Macmillan Reference, 2005.

Durbin, Paul T., ed. *Guide to Culture of Science, Technology, and Medicine*. New York: Free Press, 1984.

Durbin, Paul T., ed. *Technology and Responsibility*. Dordrecht: Reidel, 1987.

Durbin, Paul T., ed. *Broad and Narrow Interpretations of Philosophy of Technology: Philosophy and Technology* 7. Boston: Kluwer, 1990.

Eco, Umberto. *Travels in Hyperreality: Essays*. Translated by William Weaver. San Diego: Harcourt Brace Jovanovich, 1986.

Ekelund, Charlotte K., Finn Steren Jørgensen, Olav Jørn Petersen, Karin Sundberg, and Ann Tabor. "Impact of a New National Screening Policy for Down's Syndrome in Denmark: Population-Based Cohort Study." *British Medical Journal* (2008), 337:a2547.

Ellul, Jacques. *Technological Society*. Translated by John Wilkinson. New York: Vintage Books, 1964.

———. "Technological Order." In *Philosophy and Technology: Readings in the Philosophical Problems of Technology*, edited by Carl Mitcham and Robert Mackey, 86–105. New York: Free Press, 1972.

———. "The Power of Technique and the Ethics of Non-Power." In *Myths of Information: Technology and the Postindustrial Culture*, edited by Kathleen Woodward, 242–247. Madison, WI: Coda Press, 1980.

———. "The Ethics of Non-Power." In *Ethics in an Age of Pervasive Technology*, edited by Melvin Kranzberg, 204–212. Boulder, CO: Westview Press, 1980.

———. "Technology and Democracy." In *Democracy in Technological Society: Philosophy and Technology*, Vol. 9, edited by Langdon Winner, 35–50. Dordrecht: Kluwer, 1992.

Etzioni, Amitai. "Communitarianism." In *Encyclopedia of Science, Technology, and Ethics*. Detroit: Macmillan Reference, 2005.

Faden, Ruth R., and Tom L. Beauchamp. *A History and Theory of Informed Consent*. New York and Oxford: Oxford University Press, 1986.

Faulconer, James E., and Mark A. Wrathall, eds. *Appropriating Heidegger*. Cambridge: Cambridge University Press, 2000.

Feenberg, Andrew. "Subversive Rationalization: Technology, Power, and Democracy." In *Technology and the Politics of Knowledge*, edited by Andrew Feenberg and Alistair Hannay, 3–22. Bloomington and Indianapolis: Indiana University Press, 1995.

———. *Questioning Technology*. London and New York: Routledge, 1999.

———. "From Essentialism to Constructivism: Philosophy of Technology at the Crossroads." In *Technology and Good Life?*, edited by Eric Higgs, Andrew Light, and David Strong, 294–315. Chicago: University of Chicago Press, 2000.

———. "Pragmatism and Critical Theory." *Techné* 7 (2003, No. 1), 42–48.

Feenberg, Andrew, and Alistair Hannay, eds. *Technology and the Politics of Knowledge*. Bloomington and Indianapolis: Indiana University Press, 1995.

Florman, Samuel C. *Blaming Technology*. New York: St. Martin's Press, 1981.

———. *The Existential Pleasures of Engineering*. 2nd Edition. New York: St. Martin's Griffin, 1994.

Foucault, Michel. *Ethics: Subjectivity and Truth (Essential Works of Michel Foucault, 1954–1984, Vol. 1)*. New York: The New Press, 1997.

Frankena, William K. "Value and Valuation." In *The Encyclopedia of Philosophy*. New York and London: Collier Macmillan Publishers, 1967.

Frey, R. G., and Christopher Wellman, eds. *A Companion to Applied Ethics*. Malden and Oxford: Blackwell, 2003.

Fukuyama, Francis. *The End of History and the Last Man*. New York: Free Press, 1992.

———. *Our Post-Human Future: Consequences of the Biotechnology Revolution.* New York: Farrar, Straus & Giroux, 2002.

Fuller, Steve. "A Strategy for Making Science Studies Policy Relevant." In *Controversial Science: From Content to Contention*, edited by Thomas Brante, Steve Fuller, and William Lynch, 107–125. Albany: State University of New York Press, 1993.

———. "Is STS Truly Revolutionary or Merely Revolting?" *Science Studies* 18 (2005, No. 1), 75–83.

Funtowicz, S. O., J. Martinez-Alier, G. Munda, and J. R. Ravetz. *Information Tools for Environmental Policy under Conditions of Complexity.* Luxembourg: Office for Official Publications of the European Communities, 1999.

Gandhi, Mohandas K. *An Autobiography: The Story of My Experiments with Truth.* Translated by Mahadev Desai. Boston: Beacon Press, 1957.

———. *Non-Violent Resistance (Satyagraha)*, edited by Bharatan Kumarappa. New York: Dover, 2001.

Gibson, Roger F., Jr. "W. V. Quine." In *A Companion to Pragmatism*, edited by John R. Shook and Joseph Margolis, 101–107. Oxford: Blackwell, 2006.

Gill, David W. "The Technological Blind Spot in Business Ethics." *Bulletin of Science, Technology, and Society* 19 (1999, No. 3), 190–198.

———. "Business Ethics." In *Encyclopedia of Science, Technology, and Ethics*. Detroit: Macmillan Reference, 2005.

Glover, Jonathan. *Humanity: A Moral History of the Twentieth Century.* New Haven and London: Yale University Press, 2001.

Grossman, Dave. *On Killing: The Psychological Cost of Learning to Kill in War and Society.* New York and Boston: Back Bay Books, 1996.

Guignon, Charles B., ed. *The Cambridge Companion to Heidegger.* Cambridge: Cambridge University Press, 1993.

Habermas, Jürgen. *The Future of Human Nature.* Translated by William Rehg et al. Cambridge: Polity Press, 2003.

Hacking, Ian. *Representing and Intervening: Introduction Topics in the Philosophy of Natural Science.* Cambridge, New York, and Melbourne: Cambridge University Press, 1983.

Hadot, Pierre. *What Is Ancient Philosophy?* Translated by Michael Chase. Cambridge, MA: Harvard University Press, 2002.

———. *The Veil of Isis: An Essay on the History of the Idea of Nature.* Translated by Michael Chase. Cambridge, MA: Harvard University Press, 2006.

Hamamoto, Sōshun. "Master's Taste: Tea Utensils." In *Chanoyu: The Urasenke Tradition of Tea*, edited by Sōshitsu XV Sen, 52–76. Translated by Alfred Birnbaum. New York and Tokyo: Weatherhill, 1988.

Hamilton, Clive. "Downshifting in Britain: A Sea-Change in the Pursuit of Happiness." Discussion Paper 58. The Australia Institute, 2003. https://www.tai.org.au/documents/dp_fulltext/DP58.pdf

Hanks, Craig. "Review of *Philosophical Tools.*" *Techné* 7 (2004, No. 3), 130–135.

Hansson, Sven Ove. "Risk Ethics." In *Encyclopedia of Science, Technology, and Ethics*. Detroit: Macmillan Reference, 2005.

Haraway, Donna. *Simians, Cyborgs, and Women: The Reinvention of Nature.* New York: Routledge, 1991.

Hardin, Garrett. "When Technology Alters Nature, Two Kinds of Responsibility." Review of *The Imperative of Responsibility. Hastings Center Report* 14 (1984, No. 6), 45–47.

Harman, Graham. *Heidegger Explained: From Phenomenon to Thing.* Chicago and La Salle, IL: Open Court, 2007.

Harris, Charles E., Michael Pritchard, and Michael J. Rabins. "Engineering Ethics." In *Encyclopedia of Science, Technology, and Ethics*. Detroit: Macmillan Reference, 2005.

Harris, John. "Cloning." In *A Companion to Applied Ethics*, edited by R. G. Frey and Christopher Wellman, 382– 395. Malden and Oxford: Blackwell, 2003.

Harrison, Jonathan. "Ethical Naturalism." In *The Encyclopedia of Philosophy.* New York and London: Collier Macmillan Publishers, 1967.

Hautakangas, Sami, and Tomi Kiilakoski. "The Information Society: Towards an Iron Cage of E-learning?" *European Educational Research Journal* 3 (2004, No. 1), 1–13.
Hawthorne, Nathaniel. *Tales and Sketches*, edited by Roy Harvey Pearce. New York: Library of America, 1996.
Häyry, Matti. "If You Must Make Babies, Then at Least Make the Best Babies You Can?" *Human Fertility* 7 (2004a, No. 2), 105–112.
———. "A Rational Cure for Prereproductive Stress Syndrome." *Journal of Medical Ethics* 30 (2004b, No. 4), 377–378.
———. "Vastuu tulevista." Column in *Metro Helsinki*, January 5, 2006.
Heath, Joseph M. "Jürgen Habermas." In *A Companion to Pragmatism*, edited by John R. Shook and Joseph Margolis, 120–126. Oxford: Blackwell, 2006.
Hegel, Georg Wilhelm Friedrich. *Phenomenology of Spirit.* Translated by A. V. Miller. Oxford: Oxford University Press, 1977.
———. *Grundlinien der Philosophie des Rechts.* Werke 7. Frankfurt am Main: Suhrkamp, 1986.
———. *Phänomenologie des Geistes.* Werke 3. Frankfurt am Main: Suhrkamp, 1986.
———. *The Elements of the Philosophy of Right*, edited by Allen W. Wood. Translated by H. B. Hisbet. Cambridge: Cambridge University Press, 1991.
Heidegger, Martin. *Vorträge und Aufsätze.* Pfüllingen: Neske, 1959.
———. *Being and Time.* Translated by John Macquarrie and Edward Robinson. Oxford and Cambridge, MA: Blackwell, 1962.
———. *Die Technik und die Kehre.* Pfüllingen: Neske, 1962.
———. *Discourse on Thinking: A Translation of* Gelassenheit. Translated by John M. Anderson and E. Hans Freund. New York: Harper & Row, 1966.
———. *What Is Called Thinking?* Translated by J. Glenn Gray. New York: HarperCollins, 1966.
———. *Wegmarken.* Hrsg. Friedrich-Wilhelm von Hermann. Frankfurt am Main: Vittorio Klostermann, 1976.
———. *The Question Concerning Technology and Other Essays.* Translated by William Lovitt. New York: Harper & Row, 1977.
———. *Fundamental Concepts of Metaphysics.* Translated by William McNeill and Nicholas Walker. Bloomington, IN: Indiana University Press, 1995.
———. *Being and Time: A Translation of Sein and Zeit.* Translated by Joan Stambaugh. Albany: State University of New York Press, 1996.
———. *Pathmarks*, edited by William McNeill. Cambridge: Cambridge University Press, 1998.
———. *Gelassenheit.* 12 Aufl. Stuttgart: Günther Neske Verlag, 2000a.
———. *Introduction to Metaphysics.* Translated by Gregory Fried and Richard Polt. New Haven and London: Yale University Press, 2000b.
———. *Sein und Zeit.* 18. Aufl. Tübingen: Max Niemer Verlag, 2001.
———. *Poetry, Language, Thought.* Translated by Albert Hofstadter. New York: HarperCollins, 2001.
———. *Gesamtausgabe, Band 5: Holzwege.* Hrsg. Friedrich-Wilhelm von Hermann. Frankfurt am Main: Vittorio Klostermann, 2003.
Heikkerö, Topi. "The Fate of Western Civilization: G. H. von Wright's Reflections on Science, Technology and Global Society." *Bulletin of Science, Technology, and Society* 24 (2004, No. 2), 156–162.
———. "The Good Life in a Technological World: Focal Things and Practices in the West and in Japan." *Technology in Society* 27 (2005, No. 2), 251–259.
———. "Materialism." In *Encyclopedia of Science, Technology, and Ethics*. Detroit: Macmillan Reference, 2005.
———. "Ethics of Technology and the Notion of Nature: Hans Jonas's Philosophy of Life as a Grounding for His Ethics of Responsibility." In *Humanity at the Turning Point: Rethinking Nature, Culture, and Freedom. Essays on Contemporary Philosophy*, edited by Sonja Servomaa, 186–200. Helsinki: Renvall Publications 23, 2006.

———. "Ethos in Engineering Education: How to Address the Volition Dimension of the Engineer's Social Responsibility?" *European Journal of Engineering Education* 33 (2008, No. 2), 161–168.

———. "Material Culture and the Conditions for the Good Life: Albert Borgmann on Thinking about Ethics in Technological Culture." In *Praxiology and the Philosophy of Technology: The International Annual of Practical Philosophy and Methodology*, Vol. 15, edited by Timo Airaksinen and Wojciech W. Gasparski, 215–229. New Brunswick and London: Transaction Publishers, 2008.

Heinämaa, Sara. "Wonder and (Sexual) Difference: Cartesian Radicalism in Phenomenological Thinking." In *Norms and Modes of Thinking in Descartes*, edited by Tuomo Aho and Mikko Yrjönsuuri, 277–296. Helsinki: Acta Philosophica Fennica, 1999.

Hemingway, Ernest. *The Complete Short Stories of Ernest Hemingway*. New York: Scribner, 1987.

Hersey, John. *The Single Pebble*. New York: Vintage Books, 1956.

Hertzberg, Lars. "Was Wittgenstein a Moral Philosopher?" *Studia Theologica* 51 (1997, No. 2), 144–155.

———. "Moral Escapism and Applied Ethics." *Philosophical Papers* 31 (2002, No. 3), 251–270.

Hesse, Mary. "Laws and Theories." In *The Encyclopedia of Philosophy*. New York and London: Macmillan, 1967.

Hickman, Larry A. *John Dewey's Pragmatic Technology*. Bloomington, IN: Indiana University Press, 1990.

———. "Technology, Final Structures, and Focal Things." Review of *Technology and the Character of Contemporary Life*. *Research in Philosophy and Technology* 12 (1992), 337–345.

———. "Nature as Culture: John Dewey's Pragmatic Naturalism." In *Environmental Pragmatism*, edited by Andrew Light and Eric Katz, 50–72. Environmental Philosophies Series. London and New York: Routledge, 1996.

———. "Techne and Politeia Revisited: Pragmatic Paths to Technological Revolution." *Techné* 1 (1996a, No. 3–4).

———. "Four Effects of Technology." *Techné* 3 (1998, No. 4).

———. "Making the Family Functional: The Case for Legalized Same-Sex Domestic Partnerships." *Philosophy of the Social Sciences* 29 (1999, No. 2), 231–247.

———. "Focaltechnics, Pragmatechnics, and the Reform of Technology." In *Technology and Good Life?*, edited by Eric Higgs, Andrew Light, and David Strong, 89–105. Chicago: University of Chicago Press, 2000.

———. *Philosophical Tools for Technological Culture: Putting Pragmatism to Work*. Bloomington, IN: Indiana University Press, 2001.

———. "Revisiting *Philosophical Tools for Technological Culture*." *Techné* 7 (2003, No. 1), 64–81.

———. "Pragmatism, Postmodernism, and Global Citizenship." *Metaphilosophy* 35 (2004, Nos. 1/2), 65–81.

———. *Pragmatism as Post-Postmodernism: Lessons from John Dewey*. New York: Fordham University Press, 2007.

Higgs, Eric, Andrew Light, and David Strong, eds. *Technology and the Good Life?* Chicago and London: University of Chicago Press, 2000.

Himanen, Pekka. *The Hacker Ethic: A Radical Approach to the Philosophy of Business*. New York: Random House, 2001.

Himanka, Juha. "Before and After the Reduction." *The Journal of the British Society for Phenomenology* 32 (2001, No. 2), 188–204.

Homer. *The Odyssey of Homer*. Translated by Richmond Lattimore. New York: Harper and Row, 1967.

Honneth, Axel. *Kampf um Anerkennung: zur moralischen Grammatik sozialer Konflikte*. Frankfurt am Main: Suhrkamp, 1992.

———. *The Struggle for Recognition: The Moral Grammar of Social Conflicts*. Translated by Joel Anderson. Cambridge, MA: MIT Press, 1996.

Horkheimer, Max. *Eclipse of Reason*. New York: Continuum, 1992.
Hottois, Gilbert. "Technoscience." In *Encyclopedia of Science, Technology, and Ethics*. Detroit: Macmillan Reference, 2005.
Hughes, Thomas. *Human-Built World: How to Think about Technology and Culture*. Chicago and London: University of Chicago Press, 2004.
Hume, David. *A Treatise of Human Nature*, edited by L. A. Selby-Bigge. Oxford: Clarendon Press, 1958.
Husserl, Edmund. *Die Idee der Phänomenologie. Fünf Vorlesungen*. Hrsg. Walter Biemel. Husserlian II Bd. The Hague: Martinus Nijhoff, 1973.
———. *The Idea of Phenomenology*. Translated by Lee Harvey. Dordrecht: Kluwer, 1999.
Huxley, Aldous. *Brave New World*. London: Chatto & Windus, 1932.
Iannone, A. Pablo. "Pollution." In *Encyclopedia of Science, Technology, and Ethics*. Detroit: Macmillan Reference, 2005.
Ihde, Don. *Technology and the Lifeworld*. Bloomington, IN: University of Indiana Press, 1990.
———. *Postphenomenology: Essays in the Postmodern Context*. Reprinted. Evanston, IL: Northwestern University Press, 1995.
———. "Has the Philosophy of Technology Arrived? A State-of-the-Art Review." *Philosophy of Science* 71 (2004, No. 1), 117–131.
Illich, Ivan. *Tools for Conviviality*. London: Fontana, 1975.
———. *Deschooling Society*. London and New York: Marion Boyars, 1999.
Jackson, Philip W. "John Dewey." In *A Companion to Pragmatism*, edited by John R. Shook and Joseph Margolis. Oxford: Blackwell, 2006.
James, William. *Pragmatism: A New Name for Some Old Ways of Thinking: Together with Four Related Essays Selected from* The Meaning of Truth. New York and London: Longmans, Green & Co, 1949.
Jasanoff, Sheila, Gerald E. Markle, James C. Peterson, and Trevor J. Pinch, eds. *Handbook of Science and Technology Studies*. Revised Edition. Thousand Oaks, CA: Sage, 2001.
Jasper, James M. "Three Nuclear Energy Controversies." In *Controversy: Politics of Technological Decisions*, edited by Dorothy Nelkin. 3rd Edition. Newbury Park, CA: Sage Publications, 1992.
Johnson, Deborah G. *Computer Ethics*. 3rd Edition. Upper Saddle River, NJ: Prentice Hall, 2001.
———. "Computer Ethics." In *A Companion to Applied Ethics*, edited by R. G. Frey and Christopher Wellman, 608–619. Malden and Oxford: Blackwell, 2003.
———. "Responsibility: Anglo-American Perspectives." In *Encyclopedia of Science, Technology, and Ethics*. Detroit: Macmillan Reference, 2005.
Jonas, Hans. "Gnosticism and Modern Nihilism." *Social Research* 19 (1952, No. 1), 430–452.
———. *Gnostic Religion: The Message of the Alien God and the Beginnings of Christianity*. 2nd Revised Edition. Boston: Beacon Press, 1958.
———. *Gnosis und Spätantiker Geist. 1. Teil: Die Mythologische Gnosis: mit einer Einleitung zur Geschichte und Methodologie der Forschung*. Göttingen: Vandenhoeck and Ruprecht, 1964.
———. "Gnosticism." In *The Encyclopedia of Philosophy*. New York and London: Macmillan, 1967.
———. *Philosophical Essays: From Ancient Creed to Technological Man*. Englewood Cliffs, NJ: Prentice-Hall, 1974.
———. "Im Kampf um die Möglichkeit des Glaubens." In *Gedenken an Rudolf Bultmann*, edited by Otto Kaiser, 41–70. Tübingen: J. C. B. Mohr, 1977.
———. *Das Prinzip Verantwortung. Versuch einer Ethik für die technologische Zivilization*. Frankfurt am Main: Insel Verlag, 1979.
———. "Technology as a Subject for Ethics." *Social Research* 49 (1982, No. 4), 891–898.
———. *The Imperative of Responsibility: In Search of an Ethics for the Technological Age*. Translated by Hans Jonas and David Herr. Chicago and London: University of Chicago Press, 1984.
———. *Technik, Medizin und Ethik. Zur Praxis des Prinzips Verantworung*. Frankfurt am Main: Insel Verlag, 1985.

———. *Wissenschaft als persönliches Erlebnis.* Göttingen: Vandenhoeck and Ruprecht, 1987.

———. "Heidegger's Resoluteness and Resolve." In *Martin Heidegger and National Socialism: Questions and Answers*, edited by Günther Neske and Emil Kettering, 197–203. Translated by Lisa Harries and Joachim Neugroschel. New York: Paragon House, 1990.

———. *Gnosis und Spätantiker Geist. 2. Teil: Von der Mythologie zur der mystischen Philosophie.* Hrsg. Kurt Rudolph. Göttingen: Vandenhoeck and Ruprecht, 1993.

———. *Mortality and Morality: A Search for the Good after Auschwitz*, edited by Lawrence Vogel. Evanston, IL: Northwestern University Press, 1996.

———. *The Phenomenon of Life: Toward a Philosophical Biology.* Evanston, IL: Northwestern University Press, 2001. (1st Edition 1966.)

———. "*Wissenschaft* as Personal Experience." Translated by Hunter Hannum and Hildegarde Hannum. *Hastings Center Report* 32 (2002, No. 4), 27–35.

———. *Erinnerungen.* Heraussgeben von Christian Wiese. Frankfurt am Main: Insel, 2003.

———. *Memoirs*, edited by Christian Wiese. Translated by Krishna Winston. Waltham, MA: Brandeis University Press, 2008.

Jonas, Hans, and Harvey Scodel. "An Interview with Professor Hans Jonas." *Social Research* 70 (2003, No. 2), 339–368.

Jonsen, Albert. *The Birth of Bioethics.* Oxford and New York: Oxford University Press, 1998.

Joy, Bill. "Why the Future Doesn't Need Us." *Wired* 8 (2000, No. 4).

Kaiser, Otto, ed. *Gedenken an Rudolf Bultmann.* Tübingen: J. C. B. Mohr, 1977.

Kant, Immanuel. *The Grounding for the Metaphysics of Morals.* Translated by James W. Ellington. Indianapolis and Cambridge: Hackett, 1993.

———. *Grundlegung zur Metaphysik der Sitten.* Hrsg. Jen Timmermann. Göttingen: Vandenhoeck and Ruprecht, 2004.

Karnos, David D., and Robert G. Shoemaker, eds. *Falling in Love with Wisdom: American Philosophers Talk about Their Calling.* New York and Oxford: Oxford University Press, 1993.

Kass, Leon R. "Appreciating *The Phenomenon of Life.*" *Hastings Center Report* 27 (1995, No. 7), 3–12.

Kennington, Richard. *On Modern Origins: Essays in Early Modern Philosophy*, edited by Pamela Kraus and Frank Hunt. Lanham, MD: Lexington, 2004.

———. Interpretative Essay. In *Discourse on Method.* By René Descartes, edited by Pamela Krauss and Frank Hunt, 59–77. Newburyport, MA: Focus, 2007.

Kepler, Johannes. *Selections from Kepler's* Astronomia Nova, edited and translated by William H. Donahue. Santa Fe: Green Lion Press, 2008.

Keulartz, Jozef, Maartje Schermer, Michiel Korthals, and Tsjalling Swierstra. "Ethics in Technological Culture: A Programmatic Proposal for a Pragmatist Approach." *Science, Technology, and Human Values* 29 (2004, No. 1), 3–29.

Kierkegaard, Søren. *Concluding Unscientific Postscript*, Vols. 1 and 2, edited and translated by Howard V. Hong and Edna H. Hong. Princeton: Princeton University Press, 1992.

Kitching, Gavin. *Seeking Social Justice through Globalization: Escaping a Nationalist Perspective.* University Park, PA: Pennsylvania State University Press, 2001.

Kiyota, Minoru. *The Shambala Guide to Kendo: Its History, Philosophy, and Spiritual Dimension.* Boston and London: Shambala, 2002.

Klein, Jacob. *Greek Mathematical Thought and the Origin of Algebra.* Translated by Eva Brann. Cambridge, MA: MIT Press, 1968.

———. *Lectures and Essays*, edited by Robert B. Williamson and Elliot Zuckerman. Annapolis, MD: St. John's College Press, 1985.

Klein, Julie Thompson. *Interdisciplinarity: History, Theory, and Practice.* Detroit: Wayne State University Press, 1990.

———. *Crossing Boundaries: Knowledge, Disciplinarities, and Interdisciplinarities.* Charlottesville and London: University Press of Virginia, 1996.

Kojève, Alexandre. *Introduction to the Reading of Hegel. Lectures on the Phenomenology of Spirit*, edited by Raymond Queneau and Allan Bloom. Translated by James H. Nichols. Ithaca and London: Cornell University Press, 1969.

Kranzberg, Melvin, ed. *Ethics in an Age of Pervasive Technology*. Boulder, CO: Westview Press, 1980.

Kuhn, Thomas S. *The Structure of Scientific Revolutions*. 2nd Edition. Chicago and London: University of Chicago Press, 1970. (1st Edition 1962.)

———. *The Copernican Revolution: Planetary Astronomy in the Development of Western Thought*. Revised Edition. Cambridge, MA: Harvard University Press, 1985. (1st Edition 1957.)

Kundera, Milan. *The Unbearable Lightness of Being*. Translated by Michael Henry Heim. New York: HarperCollins, 1991.

Kurzweil, Ray. *The Age of Spiritual Machines: When Computers Exceed Human Intelligence*. New York: Viking, 1999.

LaFleur, William. "Infants, Paternalism, and Bioethics: Japan's Grasp of Jonas's Insistence on Intergenerational Responsibility." In *The Legacy of Hans Jonas: Judaism and the Phenomenon of Life*, edited by Hava Tirosh-Samuelson and Christian Wiese, 461–480. Leiden and Boston: Brill, 2008.

LaFollette, Hugh, ed. *The Blackwell Guide to Ethical Theory*. Malden and Oxford: Blackwell, 2000.

LaFollette, Hugh. "Pragmatic Ethics." In *The Blackwell Guide to Ethical Theory*, edited by Hugh LaFollette, 400–419. Malden and Oxford: Blackwell, 2000.

Latour, Bruno. *We Have Never Been Modern*. Translated by Catherine Porter. Cambridge, MA: Harvard University Press, 1993.

Latour, Bruno, and Steve Woolgar. *Laboratory Life: The Construction of Scientific Facts*. Revised Edition. Princeton: Princeton University Press, 1986. (1st edition 1979.)

Leibniz, Gottfried. *Philosophical Essays*, edited and translated by Roger Ariew and Daniel Garber. Indianapolis and Cambridge: Hackett, 1989.

Lenk, Hans. "Responsibility: German Perspectives." In *Encyclopedia of Science, Technology, and Ethics*. Detroit: Macmillan Reference, 2005.

Levinas, Emmanuel. *Totality and Infinity: An Essay on Exteriority*. Translated by Alphonso Lingis. Pittsburgh: Duquesne University Press, 1969.

———. "Heidegger, Gagarin, and Us." In *Difficult Freedom: Essays on*
Judaism. Translated by Sean Hand. Baltimore: The John Hopkins University Press, 1990.

Levy, David J. *Hans Jonas: The Integrity of Thinking*. Columbia and London: University of Missouri Press, 2002.

Lewis, C. S. *The Abolition of Man*. Oxford: Oxford University Press, 1943.

Liddell, George Henry, and Robert Scott. *A Greek-English Lexicon*. Revised by H. S. Jones and R. McKenzie. Oxford: Oxford University Press, 1983.

Light, Andrew. "Borgmann's *Unzeitgemässige Betrachtungen:* On the Pre-political Conditions of Politics of Place." In *Technology and the Good Life?*, edited by Eric Higgs, Andrew Light, and David Strong, 106–126. Chicago and London: University of Chicago Press, 2000.

———. "Environmental Ethics." In *A Companion to Applied Ethics*, edited by R. G. Frey and Christopher Wellman, 633–649. Malden and Oxford: Blackwell, 2003.

Light, Andrew, and Eric Katz, eds. *Environmental Pragmatism*. Environmental Philosophies Series. London and New York: Routledge, 1996.

Lindberg, Susanna. "Hans Jonas's Theory of Life in the Face of Responsibility." *Phänomenologische Forschungen*. Jahrgang 2005, 175–191. Hamburg: Felix Meiner Verlag, 2005.

Lovins, Amory B., and L. Hunter Lovins. "Brittle Times, RMI's Response." *RMI Solutions* XVII (Fall 2001, No. 3), 1–3.

Machiavelli, Niccolò. *The Prince*. Translated by Harvey Mansfield. 2nd Edition. Chicago: University of Chicago Press, 1998.

MacIntyre, Alasdair. *After Virtue: A Study in Moral Theory*. 2nd Edition. Chicago: University of Notre Dame Press, 1984.

———. "Does Applied Ethics Rest on a Mistake?" *Monist* 67 (1984, No. 4), 498–513.

Maclean, Norman. *A River Runs through It and Other Stories*. Chicago: University of Chicago Press, 1976.

Marcuse, Herbert. *One-Dimensional Man: Studies in the Ideology of Advanced Industrial Society.* Boston: Beacon Press, 1964.
Margolis, Joseph. *Reinventing Pragmatism: American Philosophy at the End of the Twentieth Century.* Ithaca, NY: Cornell University Press, 2002.
Marjanen, Antti, ed. *Was There a Gnostic Religion?* Helsinki: Finnish Exegetical Society/ Göttingen: Vandenhoeck and Ruprecht, 2005.
Marjanen, Antti. "What Is Gnosticism? From the Pastorals to Rudolph." In *Was There a Gnostic Religion?*, edited by Antti Marjanen, 1–53. Helsinki: Finnish Exegetical Society/ Göttingen: Vandenhoeck and Ruprecht, 2005.
Martin, Brian. "Technological Vulnerability." *Technology in Society* 18 (1996, No. 4), 511–523.
Martin, Mike W. "Whistle-Blowing: Professionalism, Personal Life, and Shared Responsibility for Safety in Engineering." *Business and Professional Ethics Journal* 11 (1992, No. 2), 21–37.
Martin, Mike W., and Roland Schinzinger. *Ethics in Engineering.* 4th Edition. New York: McGraw-Hill, 2005.
Marx, Karl. *Selected Writings*, edited by Lawrence H. Simon. Indianapolis: Hackett, 1994.
———. *The Poverty of Philosophy.* Elibron Classics. Boston: Adamant Media, 2005.
Marx, Leo. "Does Improved Technology Mean Progress?" *Technology Review* (January 1987), 33–41.
McDermott, John. "Technology: The Opiate of the Intellectuals." *The New York Review of Books* 13 (1969, No. 2), 25–35.
McGinn, Robert. "Technology, Demography, and the Anachronism of Traditional Rights." *Journal of Applied Philosophy* 11 (1994, No. 1), 57–70.
McGuinness, Brian, ed. *Wittgenstein and His Times.* Oxford: Blackwell, 1982.
McLemee, Scott. "Losing Friends and Influencing People." *The Chronicle of Higher Education*, January 18, 2002, A12.
McLuhan, Marshall. *Understanding Media: The Extensions of Man.* Cambridge, MA: MIT Press, 1994. (1st edition 1964.)
Merton, Robert K. *The Sociology of Science: Theoretical and Empirical Investigations*, edited by Norman W. Storer. Chicago: University of Chicago Press, 1973.
Meskus, Mianna. *Elämän tiede: Tutkimus lääketieteellisestä teknologiasta, vanhemmuudesta ja perimän hallinnasta* [Science of Life: A Study on Medical Technology, Parenthood, and the Management of Heredity]. Tampere: Vastapaino, 2009.
———. "Governing Risk through Informed Choice: Prenatal Testing in Welfarist Maternity Care." In *Contest Categories: Studies of the Life Sciences in Society*, edited by Susanne Bauer, Hanne Jensen, and Ayo Wahlberg, 49–68. Aldershot: Ashgate, 2009.
Mesthene, Emmanuel G. "The Role of Technology in Society." *Technology and Culture* 10 (1969, No. 4), 489–536.
Misak, Cheryl, ed. *Cambridge Companion to Peirce.* Cambridge and New York: Cambridge University Press, 2004.
Misak, Cheryl. "Charles Sanders Peirce 1839–1914." In *Cambridge Companion to Peirce*, edited by Cheryl Misak, 1–26. Cambridge and New York: Cambridge University Press, 2004.
Mitcham, Carl. "Philosophy of Technology." In *Guide to Culture of Science, Technology, and Medicine*, edited by Paul T. Durbin, 282–363. New York: Free Press, 1984.
———. "Schools for Whistle-Blowers: Educating Ethical Engineers." *Commonweal* 114 (1984, No. 7), 201–205.
———. "Responsibility and Technology: The Expanding Relationship." In *Technology and Responsibility*, edited by Paul T. Durbin, 3–40. Dordrecht: Reidel, 1987.
———. *Thinking through Technology: The Path between Engineering and Philosophy.* Chicago: University of Chicago Press, 1994.
———. *Thinking Ethics in Technology: Hennebach Lectures and Papers 1995–1996.* Golden, CO: Colorado School of Mines, 1997.

———. "On Character and Technology." In *Technology and the Good Life?*, edited by Eric Higgs, Andrew Light, and David Strong, 126–148. Chicago and London: University of Chicago Press, 2000.

———. Introduction. In *Encyclopedia of Science, Technology, and Ethics*. Detroit: Macmillan Reference, 2005.

———. "Nuclear Ethics: Weapons Perspectives." In *Encyclopedia of Science, Technology, and Ethics*. Detroit: Macmillan Reference, 2005.

———. "Responsibility: Overview." In *Encyclopedia of Science, Technology, and Ethics*. Detroit: Macmillan Reference, 2005.

———. "Science, Democracy, and Philosophy." In *Public Science in Liberal Democracy*, edited by Jane Porter, 40–59. Toronto: University of Toronto Press, 2007.

Mitcham, Carl, editor in chief. *Encyclopedia of Science, Technology, and Ethics*. Detroit: Macmillan Reference, 2005.

Mitcham, Carl, and Adam Briggle. "Humanization and Dehumanization." In *Encyclopedia of Science, Technology, and Ethics*. Detroit: Macmillan Reference, 2005.

Mitcham, Carl, and Robert Frodeman. "New Directions on the Philosophy of Science." *Philosophy Today* 48 (2004, No. 5, Supplement), 3–15.

Mitcham, Carl, and Robert Mackey, eds. *Philosophy and Technology: Readings in the Philosophical Problems of Technology*. New York: Free Press, 1972.

Mitcham, Carl, and Helen Nissenbaum. "Technology and Ethics." In *Routledge Encyclopedia of Philosophy*, Vol. 9, edited by E. Craig. London: Routledge, 1998.

Moor, James H. "What Is Computer Ethics?" *Metaphilosophy* 16 (1985, No. 4), 266–275.

Mulhall, Stephen. "Ethics in the Light of Wittgenstein." *Philosophical Papers* 31 (2002, No. 3), 293–321.

Mumford, Lewis. "Technics and the Nature of Man." In *Philosophy and Technology: Readings in the Philosophical Problems of Technology*, edited by Carl Mitcham and Robert Mackey, 77–85. New York: Free Press, 1972.

Murdoch, Iris. *Existentialists and Mystics: Writings on Philosophy and Literature*, edited by Peter Conradi. New York and London: Penguin, 1999.

Nakamura, Masao. "Urasenke Tearooms and Their Settings." In *Chanoyu: The Urasenke Tradition of Tea*, edited by Sōshitsu XV Sen, 96–119. Translated by Alfred Birnbaum. New York and Tokyo: Weatherhill, 1988.

Nash, Roderick Frazier. The Rights of Nature: A History of Environmental Ethics . Madison, WI: University of Wisconsin Press, 1989.

National Advisory Board on Research Ethics. *Good Scientific Practice and Procedures for Handling Misconduct and Fraud in Science*. Helsinki, 2002.

Nelkin, Dorothy, ed. *Controversy: Politics of Technological Decisions*. 3rd Edition. Newbury Park, CA: Sage Publications, 1992.

Neske, Günther, and Emil Kettering, eds. *Martin Heidegger and National Socialism: Questions and Answers*. Translated by Lisa Harries and Joachim Neugroschel. New York: Paragon House, 1990.

Nietzsche, Friedrich. *The Gay Science: With Prelude in Rhymes and an Appendix of Songs*. Translated by Walter Kaufmann. New York: Vintage Books, 1974.

———. *Thus Spoke Zarathustra*, edited by Adrian Del Caro and Robert Pippin. Translated by Adrian Del Caro. Cambridge: Cambridge University Press, 2006.

Nordmann, Alfred. *Wittgenstein's Tractatus: An Introduction*. Cambridge and New York: Cambridge University Press, 2005.

———. "Technology Naturalized: A Challenge to Design for the Human Scale." In *Philosophy and Design: From Engineering to Architecture*, edited by Pieter E. Vermaas, Peter Kroes, Andrew Light, and Steven A. Moore, 173–184. Dordrecht: Springer, 2008.

Olsen, Jan-Kyrre Berg, and Evan Selinger, ed. *Philosophy of Technology: 5 Questions*. Birkerod: Automatic Press/VIP, 2007.

Olson, Robert G. "Nihilism." *The Encyclopedia of Philosophy*. New York and London: Collier Macmillan Publishers, 1967.

Ortega y Gasset, José. *The Revolt of the Masses*. New York: New American Library of World Literature, 1950.

———. "Thoughts on Technology." In *Philosophy and Technology: Readings in the Philosophical Problems of Technology*, edited by Carl Mitcham and Robert Mackey, 290–313. New York: Free Press, 1972.

Orwell, George. *Nineteen Eighty-Four.* London: Secker & Warburg, 1955.

Ozawa, Hiroshi. *Kendo: The Definitive Guide.* New York: Kodansha International, 1997.

Passmore, John. "Logical Positivism." In *The Encyclopedia of Philosophy.* New York and London: Collier Macmillan Publishers, 1967.

Peirce, Charles S. *Writings of Charles S. Peirce: A Chronological Edition, Vol. 3: 1872–1878*, edited by C. J. W. Kloesel et al. Bloomington, IN: Indiana University Press, 1986.

Pierson, Christopher. *Beyond the Welfare State? The New Political Economy of Welfare.* 3rd Edition. Cambridge: Polity Press, 2006.

Pippin, Robert B. "On the Notion of Technology as Ideology." In *Technology and the Politics of Knowledge*, edited by Andrew Feenberg and Alistair Hannay, 43–61. Bloomington and Indianapolis: Indiana University Press, 1995.

Pirsig, Robert M. *Zen and the Art of Motorcycle Maintenance: An Inquiry into Values*. New York: William Morrow, 1974.

Plato. *Timaeus, Critias, Cleitophon, Menexenus, Epistles.* Translated by R. G. Bury. Loeb Classical Library. London: Heinemann, 1961.

———. *Republic*, I–V. Translated by Paul Shorey. Loeb Classical Library. Cambridge, MA: Harvard University Press, 1969.

———. *Republic*, VI–X. Translated by Paul Shorey. Loeb Classical Library. Cambridge, MA: Harvard University Press, 1970.

———. *Lysis, Symposium, Gorgias*. Translated by W. R. M. Lamb. Loeb Classical Library. Cambridge, MA: Harvard University Press, 1996.

———. *Euthyphro, Apology, Crito, Phaedo, Phaedrus.* Translated by Harold North Fowler. Loeb Classical Library. Cambridge, MA: Harvard University Press, 1999.

Porter, Jane, ed. *Public Science in Liberal Democracy.* Toronto: University of Toronto Press, 2007.

Power, Thomas Michael. "Trapped in Consumption: Modern Social Structure and Entrenchment of the Device." In *Technology and the Good Life?*, edited by Eric Higgs, Andrew Light, and David Strong, 271–293. Chicago and London: University of Chicago Press, 2000.

Pritchard, Michael S. "Engineering Ethics." In *A Companion to Applied Ethics*, edited by R. G. Frey and Christopher Wellman, 620–632. Malden and Oxford: Blackwell, 2003.

Prusak, Bernard G. "Cloning and Corporeality." In *The Legacy of Hans Jonas: Judaism and the Phenomenon of Life*, edited by Hava Tirosh-Samuelson and Christian Wiese, 315–344. Leiden and Boston: Brill, 2008.

Pruuki, Heli. *Kuka on potilas? Suomalaisen sikiöseulontakäytännön tarkastelua.* [Who Is the Patient? A Social-Ethical Study of the Finnish Practice of Prenatal Screening.] Helsinki: University of Helsinki, 2007.

Raittila, Hannu. *Canal Grande*. Helsinki: WSOY, 2001.

Rask, Mikko, Riikka Eela, Topi Heikkerö, and Aleksi Neuvonen. *Teknologian arviointi, arvot ja osallistuminen—kokemuksia geenitekniikka-arvioista* [Values and Participation in Technology Assessment: Experiences of Assessing Gene Technology]. Group for Technology Studies, Working Papers 45. Espoo: VTT, 1999.

Rawls, John. *Theory of Justice.* Cambridge, MA: Harvard University Press, 1971.

———. "Justice as Fairness." In *Collected Papers*, edited by Samuel Freeman, 47–72. Cambridge, MA: Harvard University Press, 1999.

Reich, Robert B. *The Future of Success: Working and Living in the New Economy.* New York: Vintage, 2000.

Restivo, Sal, ed. *Science, Technology, and Society: An Encyclopedia.* Oxford and New York: Oxford University Press, 2005.

Rothstein, Bo. *Just Institutions Matter: The Moral and Political Logic of the Universal Welfare State.* Cambridge: Cambridge University Press, 1998.

Rudolph, Kurt. "Hans Jonas and Research on Gnosticism from a Contemporary Perspective." In *The Legacy of Hans Jonas: Judaism and the Phenomenon of Life*, edited by Hava Tirosh-Samuelson and Christian Wiese, 91–106. Leiden and Boston: Brill, 2008.

Safranski, Rüdiger. *Martin Heidegger: Between Good and Evil.* Translated by Edward Osers. Cambridge, MA: Harvard University Press, 1998.

Sartre, Jean Paul. *Existentialism and Humanism.* Translated by Philip Mairet. London: Methuen, 1965.

Savulescu, Julian, and Nick Bostrom, eds. *Human Enhancement.* Oxford: Oxford University Press, 2011.

Schopenhauer, Arthur. *The World as Will and Representation,* Vol. 1. Translated by E. F. J. Payne. New York: Dover, 1969.

———. *Über die Freiheit des menschlichen Willens. Über die Grundlage der Moral. Die beiden Grundprobleme der Ethik: Behandelt in zwei akademische Preisschriften. Kleinere Schriften II.* Zürich: Diogenes Verlag, 1977.

———. *Die Welt als Wille und Vorstellung I.* Sämtliche Werke: Band I. Hrsg. Wolfgang von Löhneysen. Frankfurt am Main: Suhrkamp, 1986.

———. *On the Basis of Morality.* Translated by E. F. J. Payne. Indianapolis: Hackett, 1995.

Schot, Johan W. "Constructive Technology Assessment." In *Encyclopedia of Science, Technology, and Ethics.* Detroit: Macmillan Reference, 2005.

Schütze, Christian. "The Political and Intellectual Influence of Hans Jonas." *Hastings Center Report* 27 (1995, No. 7), 40–43.

Science and Technology Policy Council of Finland. *Science, Technology, Innovation.* Helsinki, 2006.

Sclove, Richard E. "Nuts and Bolts of Democracy: Democratic Theory and Technological Design." In *Democracy in Technological Society: Philosophy and Technology,* Vol. 9, edited by Langdon Winner, 139–157. Dordrecht: Kluwer, 1992.

———. "STS on Other Planets." In *Visions of STS: Counterpoints in Science, Technology, and Society Studies,* edited by Stephen Cutcliffe and Carl Mitcham, 111–121. Albany: State University of New York Press, 2001.

Scott, James C. *Seeing Like a State: How Certain Schemes to Improve the Human Condition Have Failed.* London and New Haven: Yale University Press, 1998.

Selinger, Evan, ed. *Postphenomenology: A Critical Guide to Ihde.* Albany: State University of New York Press, 2006.

Sen, Sōshitsu XV, ed. *Chanoyu: The Urasenke Tradition of Tea.* Translated by Alfred Birnbaum. New York and Tokyo: Weatherhill, 1988.

Servomaa, Sonja, ed. *Humanity at the Turning Point: Rethinking Nature, Culture, and Freedom. Essays on Contemporary Philosophy.* Helsinki: Renvall Publications, 2006.

Shelley, Mary. *Frankenstein, or the Modern Prometheus.* In *Three Gothic Novels,* edited by Peter Fairclough. Harmondsworth, UK: Penguin Books, 1968.

Shook, John R., and Joseph Margolis, eds. *A Companion to Pragmatism.* Oxford: Blackwell, 2006.

Shrader-Frechette, Kristin S. *Nuclear Power and Public Policy: The Social and Ethical Problems of Fission Technology.* Dordrecht, London, and Boston: Reidel, 1980.

———. *Risk Analysis and Scientific Method: Methodological and Ethical Problems with Evaluating Societal Hazards.* Dordrecht, Boston, and Lancaster: Reidel, 1985.

———. *Risk and Rationality: Philosophical Foundations for Public Reforms.* Berkeley, Los Angeles, and Oxford: University of California Press, 1991.

———. "Technology, Bayesian Policymaking, and Democratic Process." In *Democracy in Technological Society: Philosophy and Technology,* Vol. 9, edited by Langdon Winner, 123–138. Dordrecht: Kluwer, 1992.

———. *Ethics of Scientific Research.* Lanham and London: Rowman & Littlefield, 1994.

———. "Technology Assessment." In *Encyclopedia of Bioethics.* 3rd Edition. New York: Macmillan, 1995.

———. *Environmental Justice: Creating Equality, Reclaiming Democracy.* Oxford and New York: Oxford University Press, 2002.

Snow, C. P. *The Two Cultures.* Cambridge: Cambridge University Press, 1993.

Soho, Takuan. *The Unfettered Mind: Writings of the Zen Master to the Sword Master.* Translated by William Scott Wilson. Tokyo, New York, and London: Kodansha International, 1986.

Son, Wha-Chul. "Autonomous Technology." In *Encyclopedia of Science, Technology, and Ethics*. Detroit: Macmillan Reference, 2005.

———. *Modern Technology and Democracy*. PhD Dissertation. Leuven: Katholieke Universiteit Leuven, 2005.

———. "Philosophy of Technology and Macro-ethics in Engineering." *Science and Engineering Ethics* 14 (2008, No. 3), 405–415.

Spengler, Oswald. *The Decline of the West: Volume I, Form and Actuality*. Translated by Charles Francis Atkinson. New York: Alfred A. Knopf, 1945.

———. *The Decline of the West: Volume II, Perspectives of World History*. Translated by Charles Francis Atkinson. New York: Alfred A. Knopf, 1945.

Spier, Raymond E. "Nuclear Ethics: Industrial Perspectives." In *Encyclopedia of Science, Technology, and Ethics*. Detroit: Macmillan Reference, 2005.

Stern, David G. *Wittgenstein's Philosophical Investigations: An Introduction*. Cambridge and New York: Cambridge University Press, 2002.

Strauss, Leo. "The Three Waves of Modernity." In *Political Philosophy: Six Essays by Leo Strauss*, edited by Hilail Gildin, 81–98. Indianapolis: Bobbs-Merrill, 1975.

———. *The Rebirth of Classical Political Rationalism: An Introduction to the Thought of Leo Strauss*, edited by Thomas Pangle. Chicago and London: University of Chicago Press, 1989.

Strong, David. *Crazy Mountains: Learning from the Wilderness to Weigh Technology*. Albany: State University of New York Press, 1995.

———. "Philosophy in the Service of Things." In *Technology and the Good Life?*, edited by Eric Higgs, Andrew Light, and David Strong, 316–338. Chicago and London: University of Chicago Press, 2000.

———. "Environmental Ethics." In *Encyclopedia of Science, Technology, and Ethics*. Detroit: Macmillan Reference, 2005.

Strong, David, and Eric Higgs. "Albert Borgmann's Philosophy of Technology." In *Technology and the Good Life?*, edited by Eric Higgs, Andrew Light, and David Strong, 19–37. Chicago and London: University of Chicago Press, 2000.

Suzuki, Daisetz Teitaro. *Zen Buddhism: Selected Writings of D.T. Suzuki*. New York: Doubleday, 1956.

———. *Zen and Japanese Culture*. Princeton: Princeton University Press, 1959.

Swierstra, Tsjalling, and Jaap Jelsma. "Responsibility without Moralism in Technoscientific Design Practice." *Science, Technology, and Human Values* 31 (2006, No. 3), 309–332.

Talosela, Minttu. *Hajatelmia bioetiikasta* [Fragmentary Reflections on Bioethics]. MTh Thesis. University of Helsinki: The Library of Theological Faculty, 2005.

Taylor, Charles. *Hegel*. Cambridge: Cambridge University Press, 1975.

———. *Ethics of Authenticity*. Cambridge, MA: Harvard University Press, 1992.

Teich, Albert, ed. *Technology and the Future*. 9th Edition. Belmont, CA: Wadsworth/Thompson, 2003.

Thayer, H. S. "Pragmatism." In *The Encyclopedia of Philosophy*. New York and London: Collier Macmillan Publishers, 1967.

Thompson, Caleb. "Wittgenstein, Tolstoy and the Meaning of Life." *Philosophical Investigations* 20, (1997, No. 2), 97–116.

Thompson, Paul T. *The Spirit of the Soil: Agriculture and Environmental Ethics:* New York: Routledge, 1995.

———. "Farming as Focal Practice." In *Technology and the Good Life?*, edited by Eric Higgs, Andrew Light, and David Strong, 166–181. Chicago and London: University of Chicago Press, 2000.

———. "Putting Pragmatism to Work?" *Techné* 7 (2003, No. 1), 59–63.

———. "Ihde and Technological Ethics." In *Postphenomenology: A Critical Guide to Ihde*, edited by Evan Selinger, 109–116. Albany: State University of New York Press, 2006.

Thoreau, Henry David. *The Portable Thoreau*, edited by Carl Bode. New York: Viking Press, 1964.

Tirosh-Samuelson, Hava, and Christian Wiese, eds. *The Legacy of Hans Jonas: Judaism and the Phenomenon of Life*. Leiden and Boston: Brill, 2008.

Tolstoy, Leo. *Death of Ivan Ilyich and Other Stories.* Translated by Aylmer Maude and J. D. Duff. New York: Penguin, 2003.

Toulmin, Stephen. "How Medicine Saved the Life of Ethics." In *New Directions in Ethics: The Challenge of Applied Ethics*, edited by Joseph P. DeMarco and Richard M. Fox, 265–281. New York and London: Routledge and Kegan Paul, 1986.

Tuchman, Barbara W. *The Guns of August.* New York: Macmillan, 1962.

Tuman, Myron. "Holding On, and Letting Go: A Review of *Holding On to Reality: The Nature of Information at the Turn of the Millennium.*" *Techné* 6 (2002, No. 1), 15–27.

Turgenev, Ivan. *Fathers and Sons.* Translated by Richard Freeborn. Oxford and New York: Oxford University Press, 1998.

Vadén, Tere. "Ethics and Gods: How Is Local Ethics Possible?" *Continental Philosophy Review* 37 (2005, No. 4), 407–437.

Vanderburg, Willem H. "Preventive Engineering: Strategy for Dealing with Negative Social and Environmental Implications of Technology." *Journal of Professional Issues in Engineering Education and Practice* 121 (1995, No. 3), 155–160.

Verbeek, Peter-Paul. *What Things Do: Philosophical Reflections of Technology, Agency, and Design.* University Park, PA: The Pennsylvania State University Press, 2005.

———. "Obstetric Ultrasound and the Technological Mediation of Morality: A Postphenomenological Analysis." In *Human Studies* 31 (2008, No. 1), 11–26.

Vermaas, Pieter E., Peter Kroes, Andrew Light, and Steven A. Moore, eds. *Philosophy and Design: From Engineering to Architecture*. Dordrecht: Springer, 2008.

Vlastos, Gregory. *Socrates: Ironist and Moral Philosopher.* Ithaca, NY: Cornell University Press, 1991.

Vogel, Lawrence. "Hans Jonas's Diagnosis of Nihilism: The Case of Heidegger." *International Journal of Philosophical Studies* 3 (1995a, No. 1), 55–72.

———. "Does Environmental Ethics Need a Metaphysical Grounding?" *Hastings Center Report* 25 (1995b, No. 7), 30–39.

———. "Hans Jonas's Exodus: From German Existentialism to Post-Holocaust Theology." Editor's Introduction. In *Mortality and Morality: A Search for the Good after Auschwitz*, edited by Lawrence Vogel, 1–40. Evanston, IL: Northwestern University Press, 1996.

———. Preface. In *The Phenomenon of Life: Toward a Philosophical Biology*, xi–xxi. Evanston, IL: Northwestern University Press, 2001.

———. "Natural Law Judaism? The Genesis of Bioethics in Hans Jonas, Leo Strauss, and Leon Kass." *Hastings Center Report* 36 (2006, No. 3), 32–44.

Vogel, Steven. "New Science, New Nature: The Habermas-Marcuse Debate Revisited." In *Technology and the Politics of Knowledge*, edited by Andrew Feenberg and Alistair Hannay, 23–42. Bloomington and Indianapolis: Indiana University Press, 1995.

Waldstein, Michael. "Hans Jonas' Construct 'Gnosticism': Analysis and Critique." *Journal of Early Christian Studies* 8 (2000, No. 3), 341–372.

Weber, Max. *From Max Weber: Essays in Sociology*, edited and translated by H. H. Gerth and C. Wright Mills. London: Routledge and Kegan Paul, 1948.

———. *Gesamtausgabe.* Abteilung I, Band. 17. ("Wissenschaft als Beruf," "Politik als Beruf.") Hrsg. Wolfgang J. Mommsen et al. Tübingen: Mohr, 1992.

White, Lynn, Jr. *Machina ex Deo: Essays in the Dynamism of Western Culture.* Cambridge, MA: MIT Press, 1968.

Wiese, Christian. *The Life and Thought of Hans Jonas: Jewish Dimensions.* Translated by Jeffrey Grossman and Christian Wiese. Waltham, MA: Brandeis University Press, 2007.

Wilshire, Bruce. *The Primal Roots of American Philosophy: Pragmatism, Phenomenology, and Native American Thought.* University Park, PA: Pennsylvania State University Press, 2000.

Winner, Langdon. *Autonomous Technology: Technics-out-of-Control as a Theme in Political Thought.* Cambridge, MA, and London: MIT Press, 1977.

———. *The Whale and the Reactor: A Search for Limits in an Age of High Technology.* Chicago and London: University of Chicago Press, 1986.

———. "Engineering Ethics and Political Imagination." In *Broad and Narrow Interpretations of Philosophy of Technology.* Philosophy and Technology, Vol. 7, edited by Paul T. Durbin, 53–64. Boston: Kluwer, 1990.

———. "Upon Opening the Black Box and Finding It Empty: Social Constructivism and the Philosophy of Technology." *Science, Technology, and Human Values* 18 (1993, No. 3), 262–378.

———. "Citizen Virtues in a Technological Order." In *Technology and the Politics of Knowledge*, edited by Andrew Feenberg and Alistair Hannay, 65–84. Bloomington and Indianapolis: Indiana University Press, 1995.

———. "Are Humans Obsolete?" *The Hedgehog Review* (Fall 2002), 25–44.

Winner, Langdon, ed. *Democracy in Technological Society: Philosophy and Technology*, Vol. 9. Dordrecht: Kluwer, 1992.

Wittgenstein, Ludwig. *Tractatus Logico-Philosophicus.* Translated by C. K. Ogden. German-English Edition. London: Kegan Paul, 1922.

———. "A Lecture on Ethics." *Philosophical Review* 74 (1965, No. 1), 3–12.

———. *Vermischte Bemerkungen.* Herausgegeben von G. H. von Wright and Heikki Nyman. Frankfurt am Main: Suhrkamp, 1977.

———. *Culture and Value*, edited by G. H. von Wright and Heikki Nyman. Translated by Peter Winch. Chicago: University of Chicago Press, 1980.

———. *Philosophical Investigations.* English-German Edition. Translated by G. E. M. Anscombe. Oxford: Blackwell, 1997.

———. *Tractatus Logico-Philosophicus.* Translated by David Pears and Brian McGuinness. London and New York: Routledge, 2001.

———. *On Certainty*, edited by G. E. M. Anscombe and G. H. von Wright. Translated by Denis Paul and G. E. M. Anscombe. Oxford: Blackwell, 2003.

Wolf, Clark. "Intergenerational Justice." In *A Companion to Applied Ethics*, edited by R. G. Frey and Christopher Wellman, 279–294. Malden and Oxford: Blackwell, 2003.

Wolin, Richard. *Heidegger's Children: Hannah Arendt, Karl Löwith, Hans Jonas, and Herbert Marcuse.* Princeton and Oxford: Princeton University Press, 2001.

———. "Ethics after Auschwitz: Hans Jonas's Notion of Responsibility in a Technological Age." In *The Legacy of Hans Jonas: Judaism and the Phenomenon of Life*, edited by Hava Tirosh-Samuelson and Christian Wiese, 1–15. Leiden and Boston: Brill, 2008.

Woodward, Kathleen, ed. *Myths of Information: Technology and the Postindustrial Culture.* Madison, WI: Coda Press, 1980.

Von Wright, Georg Henrik. "Wittgenstein in Relation to His Times." In *Wittgenstein and His Times*, edited by Brian McGuinness, 108–120. Oxford: Blackwell, 1982.

———. *Vetenskapen och förnuftet. Ett försök till orientering* [Science and Reason: An Attempt at Orientation]. Borgå: Söderström, 1986.

———. *Ciencia y razón: una tentativa de orientación.* Translated by Roberto J. Vernengo and Ricardo González. Valparaíso: Universidad de Valparaíso, 1995.

———. *Tieto ja ymmärrys.* Helsinki: Otava, 1999.

Wynne, Brian. "Public Understanding of Science." In *Handbook of Science and Technology Studies*, edited by Sheila Jasanoff, Gerald E. Markle, James C. Peterson, and Trevor J. Pinch, 361–387. Revised Edition. Thousand Oaks, CA: Sage, 2001.

Zahavi, Dan. *Husserl's Phenomenology.* Stanford, CA: Stanford University Press, 2003.

Zuboff, Shoshana. *In the Age of the Smart Machine: The Future of Work and Power.* New York: Basic Books, 1988.

Index

accident, 16, 17, 43n20, 44n36, 46n67
Accident Investigation Board Finland, 46n67
acrasia (weakness of the will), 117
Adams, Henry, 48n93
Aeschylus, 89n103
affluence, 70, 114
agency, 3, 15, 16, 23, 26, 28, 29, 31, 40, 41, 97, 184, 188, 206
Ainslie, Donald C., 43n21, 48n97, 198n20
Almond, Brenda, 13n16, 48n95
analytic philosophy. *See* philosophy, analytic
Anerkennung. *See* recognition
Anscombe, G. E. M., 47n84, 48n94
anthropocentrism, 18, 31, 59, 71, 77–78
anthropology, 18, 28–29, 36, 38, 53, 66; Albert Borgmann's, 117–119; Hans Jonas on, 58, 72–73, 74, 76, 78–79, 90n115, 90n116
anthropomorphism, 59
Apel, Karl-Otto, 31, 47n78, 90n111, 140
Arendt, Hannah, 39, 49n113, 50n117, 52, 83n2
Aristotle, 46n73, 62, 111, 117, 125n9, 138, 197n13, 200n42; Dewey's criticism of, 147; on pleasure, 117; on the soul, 60, 86n43; on praxis-poiesis distinction, 29, 171n91; on virtue, 191; on ways of life, 185, 195
Armstrong, Susan J., 43n25
Arnhart, Larry, 46n68
art, 4, 113, 122, 123, 179, 203
arts and crafts, 110
askesis, 192
Attfield, Robin, 43n25, 43n26, 43n27
authenticity, 11, 45n54, 57, 85n29, 112, 118, 119, 121, 179–180; Heidegger on, 36, 48n98, 56, 66, 84n20, 85n27, 85n29, 89n98, 112; Jean-Paul Sartre on, 85n29; Charles Taylor on, 121, 128n69, 131n130, 132n142
autonomy, 23, 77, 163, 167, 173n140, 187; in steering technological development, 24; Kant on, 131n125; in reproduction, 186, 198n24
availability, 97, 102, 103–104
Ayer, A. J., on ethics, 32, 47n80, 47n81, 47n82, 47n83

Bacon, Francis, 44n50, 88n84, 100, 127n32, 175, 191, 194, 199n33
Battin, Margaret P., 43n28, 198n16
Beauchamp, Tom L., 44n41, 48n95, 48n96, 48n97, 198n20
Being, 80, 177; and authenticity, 84n20; Heidegger on, 5, 36–38, 86n41, 159; and good (Jonas on), 61–67, 77
Bernstein, Richard, 75, 83n2, 83n7, 84n9, 90n110
Bhopal catastrophe, 16
Bijker, Wiebe, 2, 12n4, 46n76

Bland, Kalman P., 84n17
Bloch, Ernst, 69
Bloom, Allan, 156
body, 55, 56, 65, 73, 80, 84n11, 85n34, 104, 110, 124, 147, 149, 193, 202
boredom, 104
Borgmann, Albert, 7–8, 11, 94, 44n45, 45n55, 46n74; on Heidegger and ethics, 48n104; on science, 7, 98–99, 116–117
Bostrom, Nick, 46n70
Botzler, Richard G., 43n25
Brandt, Richard B., 47n82
Briggle, Adam, 46n72
Brittan, Gordon, 127n41
Bronowski, Jacob, 2, 12n3, 198n15
Buber, Martin, 118, 131n126
Bugbee, Henry G., 125n5
Bultmann, Rudolf, 53
business ethics. *See* ethics, business
Byerly, Radford, Jr., 43n20

Callahan, Daniel, 42n11, 43n27
Callon, Michel, 2
Campanella,Tommaso, 100
Carson, Rachel, 2, 42n4, 43n33
casuistry, 18, 35, 198n20
Catholic Church, 94, 117
chādo, 123–124
Challenger, space shuttle, 18, 43n20
character, 98–99, 122–123
Chernobyl, power plant accident, 16
class division, 20, 107, 123, 155
cloning, 10, 72, 84n10, 89n98
Collins, Harry, 2, 43n20, 46n76, 90n109, 126n18
Committee on Science, Engineering, and Public Policy (U.S.), 42n9, 199n29
commodification, 130n109
commodity, 102–103
community, 6, 66, 76, 104, 124, 136, 202
compact disc (CD), 102–103, 104
compassion (*Mitleid*), 161, 192, 200n41, 204
computer (personal), 4; games, 113, 120; programming, 93, 113
consciousness, 58, 59, 64, 77, 164, 202; moral, 57, 204; of mortality, 73; and recognition, 119
consumption (consumer), 22, 25, 103–104, 111, 117, 127n51, 176, 199n31
contemplation, 4, 144, 157, 197n5
control, 111, 125n11, 144, 151, 159–160, 187, 192, 194, 195, 199n38, 204; behavioral, 54; of the formation of habits, 171n114
Copernicus, Nicolaus, 54, 175, 196n1
Coplestone, Frederick, 168n17
Corrado, Michael L., 47n84, 48n94
cost-benefit analysis, 20
Crawford, Matthew B., 199n28
curriculum, 161, 189
Cutcliffe, Stephen H., 12n4
cyborg, 28, 46n65, 46n66
C-5A whistle-blowing incident, 17

Darwall, Stephen, 48n94, 48n97
Dasein (human existence), 36–38, 48n98, 58, 66, 85n27, 85n29, 86n41, 112
Das Man (the subject of inauthentic existence), 36, 45n54, 48n98, 85n27
Davis, Michael, 42n3, 42n13, 43n16, 43n19, 43n20
DC-10 accident, 17
De George, Richard T., 43n17
death, 19, 56, 59, 60, 61, 86n43; brain death, 72–73, 84n10, 89n101; definition of, 19, 89n101; human finitude and, 56, 73, 85n27, 89n103, 90n116, 112, 123, 179. *See also* mortality
decontextualization, 102
democracy, 11, 22, 24–25, 45n51, 73, 107, 147, 161, 163, 167; and the end of history, 131n131; and the minimal task of ethics, 189–190; and technology, 9, 100; participatory, 150; problems of, 89n106
Descartes, René, 44n50, 59, 84n11, 84n14, 88n84, 100, 127n32, 175, 194, 197n2, 199n34; on generosity (*generosité*), 191; on God, 118; on mastery of nature, 24–25
detachment. *See Gelassenheit*
device, 1, 5; combining with a human-being, 28
the device paradigm, 11, 94, 99–100, 102–105, 106–108, 109–111, 112, 114–115, 127n51, 182, 202; and

choices, 165; and education, 129n71; and environmental awareness, 191; and human weaknesses, 117–118; and meso-level ethics, 205; and polarities, 103; and shopping malls, 130n109; and values, 106, 177–178. *See also* discourse, paradeictic
Dewey, John, 2, 11–12, 129n79, 137, 138–139, 140; on accommodation, 143; on democracy, 161; on education, 161; and experimentation, 171n92; on habits, 157–158, 185; on individuality, 148; on Inquiry, 142, 144; and instrumental reason, 162–164; and method, 169n44; on morality, 158, 161, 172n119; on nature, 147, 169n34; on perception, 156; on philosophy, 145; and philosophy of technology, 162; on religion, 160; on thinking, 153
Didier, Christelle, 42n15
disburdenment, 102–104, 109, 117, 123, 202
discourse, 2, 27, 98–99, 204; apodeictic, 98; deictic, 98, 111, 122; ethics as, 6, 8, 126n24, 184, 201; ethics of, 76; legal, 27; and the Other, 118, 197n11; paradeictic, 98–99; political, 31; public, 115, 185; scientific, 106; taken over by the device paradigm, 108
disembodiment, 104
disengagement, 45n53, 102, 104
disposability, 102, 109. *See also* availability; commodification; device paradigm
Dōgen, Eihei, 123, 133n151
doing. *See praxis*
Dostoevsky, Fyodor, 49n106
downshifting, 195
Down syndrome, 185–186, 198n23, 199n25
Downey, Gary, 42n14
Dreyfus, Hubert L., 13n12, 49n107, 49n109, 128n64, 132n145
dualism, 10, 55–58, 61, 70, 72–73, 74, 86n43, 147, 181, 193, 203
Dubreuil, Bertrand Hériard, 42n15
Durbin, Paul T., 140, 168n13, 168n21
eating, 58, 64, 85n34, 110, 117; culture of the table, 110, 118, 121
Eco, Umberto, 128n60
ecology, 3, 16, 18, 25, 28, 47n77, 53, 90n108, 114, 154, 176; and democracy, 89n106; and the future generations, 51; and the good life, 45n53; and the human willing, 79; and naturalized technology, 144; and way of life, 195
education, 3, 29, 161, 189–190, 195; and citizenship, 114, 156, 160; commodified, 103; engineering, 3, 16, 200n45; and information technology, 111, 129n71; sex, 155–156; and the sophists, 48n99
efficiency, 100, 172n119, 184, 186, 187, 192, 199n38; as an end in itself, 152
Eigentlichkeit. *See* authencity
Ellul, Jacques, 2, 41, 44n45, 46n69, 50n117, 81, 96, 128n68, 152; on ethics of non-power, 200n44; on hope, 159–160, 172n122
ends (and means), 44n45, 57, 63–65, 71, 95–97, 99, 103, 106, 124, 129n72, 136, 139, 151–152, 162, 178, 182, 202; in focal practices, 110, 117; separated only abstractly, 149
engineering, 16–17; history of, 16–17. *See also* education, engineering; ethics, engineering
entertainment, 103–104, 128n56
environment, 18, 24, 28, 76, 139, 143, 164, 178, 191, 193
environmental pollution, 16, 43n33, 176
ethics, 6–7; applied, 8, 9, 10, 13n16, 16, 18, 20, 35, 39–41, 48n99, 51, 184, 188, 201, 205–206; bioethics, 7, 10, 16, 19–20, 22, 40, 42n10, 42n11, 43n29, 87n52, 145, 184, 189, 193, 198n16, 199n30, 201; business, 10, 16, 42n10, 189; computer, 10, 16; emotive theory of, 32; feminist, 18; engineering, 8, 10, 16–18, 25, 27, 42n10, 158, 184, 189, 201; environmental, 10, 16, 18–19, 20, 40, 42n10; food, 16; foundations of, 8, 10, 41, 29–34, 36–38, 47n78, 51–52, 53, 74, 76, 118, 192, 201–202; intergenerational, 8, 28, 51, 72, 82; macro-level, 55–56, 88n88, 185–187,

201, 205; medical, 8, 43n29; metaethics, 19, 32, 35, 176, 184; micro-level, 12, 99, 185–187, 205; mid-level, 12, 182, 184–188, 190, 195, 198n20, 199n27, 203, 205–206; normative, 6, 16, 32, 40, 184, 189, 201; nuclear, 16; of technology, 7, 9, 11–12, 39, 39–40, 81–82, 136–137, 182–196, 203–206; traditional, 26–31, 29, 54, 199n38; virtue, 18, 31, 48n94, 185, 205. *See also* principlism
evil, 79–81
excellence, 8, 110, 114, 122, 189; ideal of, 101, 127n38
expertise, 22, 35
explanation, 95–97, 126n26; and life, 59; paradigmatic, 98–99, 126n25; scientific, 95, 116; teleological, 54

fact, 88n69, 156, 159, 178, 179, 183
fact-value distinction, 136, 147, 156–157, 178, 204
Faden, Ruth R., 44n41
family, 79, 102, 109, 110, 114, 184
farming, 120, 132n138, 132n139
fate, 55, 67
Feenberg, Andrew, 9, 44n48, 119, 126n18, 131n132, 132n133, 132n134, 150, 170n70, 198n17; critiquing Hickman, 163–164, 170n70, 173n145, 173n146, 173n147
feminism, 18, 46n66, 190, 199n31
Ferré, Frederick, 126n28
Florman, Samuel C., 43n23
fly-fishing, 110, 113, 120
focal things and practices, 202; comparisons among, 113; description of, 109–111; in other cultures, 122; in Japan, 122–124; and temporality, 111; testability, 120–122; in weighing technology, 113; and willing, 193. *See also* discourse, deictic
form of life (*Lebensform*), 70; technological, 5, 35, 39, 195; Wittgenstein on, 5, 33–34
Foucault, Michel, 50n117, 143, 198n22, 205, 206n1
Friendship, 119
Frodeman, Robert, 42n9, 90n109
Fukushima nuclear power plant catastrophe, 44n36
Fukuyama, Francis, 46n71, 88n88, 131n130, 131n131
Fuller, R. Buckminster, 127n51
Fuller, Steve, 50n114
Fundamentalism, 149, 155, 160, 171n99
Funtowicz, S. O., 43n35
future generations, 8, 28, 51, 82

Galbraith, John Kenneth, 114, 130n111
Galilei, Galileo, 4, 54, 175
Gandhi, Mohandas Karamdchand, 200n41, 200n43
Gardening, 123
Gelassenheit, 80–81, 159, 195, 199n39
Gibson, Roger F. Jr., 168n18
Gill, David W., 42n10
Globalization, 20, 43n31
Glover, Jonathan, 42n1, 46n62, 148, 170n56
Gnosticism, 52, 55–57, 80–81, 84n16, 84n17, 84n18, 84n19, 87n61
God: call of, 121; as creator, 63; death of, 37; Descartes's notion of, 118; and disenchantment (Max Weber on), 179; Jacques Ellul on, 159; existence of, 71, 87n52; Gnostic, 55–56; Hans Jonas's notion of, 53; and responsibility, 27
gods, 108, 131n128, 179; Heidegger on (a god), 159
good, 83n3, 176–177; and being, 176–177; Hegel on, 128n69; Hans Jonas's criticism of contemporary perception of, 88n87; In-itself, 65, 163. *See also* *summum bonum*
the good life, 94, 101, 112, 122; contemporary conception of the good life, 88n87; and ideologies, 114, 115; and liberal democracy, 97, 107–108; and material culture, 184; and technology, 8, 24–25, 45n53–45n54, 100–101, 105, 109, 124, 178; and things, 111
Grossman, Dave, 42n1

Habermas, Jürgen, 46n71, 88n88, 136, 140, 150, 168n18

habituation, 182; habit, 137, 157, 169n29, 185; habitual bahavior, 139, 142, 143, 153, 157
hacker ethics, 104
Hadot, Pierre, 198n21, 199n39
happiness, 79, 80, 101, 107, 127n41, 179. *See also* virtue; well-being
Haraway, Donna, 46n66
harm, 15, 20, 26, 31
Harman, Graham, 84n20
Harris, John, 89n98
Hautakangas, Sami, 129n71
Hawthorne, Nathaniel, 127n44
Hebrew Scriptures, 87n60, 131n128, 159, 155, 197n4
Hegel, G. W. F., 62, 67, 86n43, 131n130, 131n131
Heidegger, Martin, 2, 6, 32, 36–39, 44n45, 52–53, 56, 57, 81, 90n116, 94, 193, 203, 206; *The Age of the World Picture*, 37–38, 49n108, 90n109, 91n125; on authenticity (*Eigentlichkeit*), 112; *Being and Time*, 36, 38, 45n54, 48n98, 52, 84n20, 85n27, 86n41, 140, 197n10; on conscience, 48n98; and dualism, 56; on death, 85n27; on everyday existence, 112, 168n15; on fallenness (*Verfallenheit*), 197n10; on finitude of human existence, 56; on *Gelassenheit* (detachment), 80–81; Larry Hickman's criticism of, 150, 159–160, 172n122; Hans Jonas on, 83n4, 83n5, 85n26; on inauthencity, 45n54; *Letter on Humanism*, 49n105; on life, 86n41; on "originary ethics," 49n105; and the other person, 118; *The Question Concerning Technology*, 38, 26, 49n112, 129n77, 199n39; as rejective of moral philosophy, 45n54, 48n104, 85n29; on "resoluteness" (*Entschlossenheit*), 57; on things, 109, 130n118, 132n150; on values, 37–38, 128n64
Heinämaa, Sara, 55
Helsinki, 130n106
Hemingway, Ernest, 129n82
Hesse, Mary, 125n11
Hertzberg, Lars, 130n94, 130n96, 130n99
heuristics of fear, 111, 196
Hickman, Larry A., 7–8, 11–12, 125; criticized by Borgmann, 162–163; criticized by Feenberg, 163–164; criticizing Borgmann, 119–120; on education, 155–156, 160–161; on ethics and technology, 42n2, 157–159; on facts and values, 156–157, 178–179, 181, 183; on habituation, 185; naturalizing technology, 141–145; on objectification, 129n79; philosophical position, 136–137, 140–141; on pragmatism, 149–151; on *praxis, poiesis*, and *theoria*, 146–149; on the public engagement of scientists, 160; on reform of technology, 159–162, 182; on religion, 149, 155, 160, 171n99; on technology and social change, 44n45, 151–153; on technoscience, 13n7
Higgs, Eric, 125n5, 126n21, 126n23, 126n24
Himanka, Juha, 130n117
Himanen, Pekka, 127n51
Hippocratic Oath, 19
history: end of, 131n130, 131n131; of technology, 2, 99–100
Hobbes, Thomas, 175
Hobbies, 120–121, 132n138
Homer, 131n128
Honneth, Axel, 131n130
hope, 159–160
Horkheimer, Max, 150, 162, 172n136
hospitality, 131n128
Hottois, Gilbert, 13n7
Hughes, Thomas, 45n51
human being (man). *See* anthropology
Hume, David, 62, 87n52, 138
Husserl, Edmund, 50n117, 85n34, 90n109, 94, 116, 130n117
Huxley, Aldous, 128n56
hybrid, 2, 13n5, 6, 90n123, 198n24
hyperactivity, 111
hyperreality, 105, 111, 128n60, 129n93
Häyry, Matti, 83n1

the idea of Man, 71–72, 75–76
ideology, 115, 122, 132n138, 150, 179–181
Ihde, Don, 13n15, 46n67, 136, 140, 168n3, 168n20

Illich, Ivan, 2
infant, 67, 78, 88n78
informed consent, 22, 44n41, 44n42
instrumentalism: as Dewey's position, 138–139, 171n92; view of science, 125n11; view of technology, 44n45, 97, 149, 151, 152
insurance policy, 103, 114
interdisciplinarity, 9, 39–41, 188, 201, 204, 206

James, William, 137, 138, 157, 168n10
Japan, 43n26; Japanese culture, 122–124; Japan Science and Technology Agency, 21, 190. *See also* Fukushima
Jasper, James M., 44n47
Jelsma, Jaap, 13n17
Jesus of Nazareth, 155, 194; Sermon on the Mount, 185, 131n128
Johnson, Deborah G., 13n14
Jonas, Hans: on brain-death, 72–73, 84n10; on cloning, 72, 84n10, 89n98; on democracy, 89n106; *Gnostic Religion*, 53, 84n16, 84n17; on Gnosticism, 52, 55–57, 80, 84n16–84n19, 87n61; on Heidegger, 52, 56, 57, 83n2, 83n4, 83n5, 85n26, 87n51; on hobbies, 132n138; *Imperative of Responsibility*, 10, 53, 63–65, 67–73, 75–77, 81, 202; life and work, 52–53; on metaphysics, 60, 61–62, 73, 74–75, 80, 85n30, 87n51, 87n53, 87n58, 87n59, 181; on utopianism, 68–69, 75–76, 88n88; *Phenomenon of Life*, 10, 53, 59–61, 86n37. *See also* nihilism
Jonsen, Albert, 43n28
Joy, Bill, 88n88
justice, 6, 163, 167, 185, 201, 203, 205; and technology, 20; and human survival, 76; and the Other, 197n11; participatory, 22; social justice, 190

Kant, Immanuel, 26, 45n58, 62, 125n9, 187
Kapp, Ernst, 2
Kass, Leon R., 78, 90n116
kendō, 122–123, 124, 132n146, 132n150
Kennington, Richard, 84n11, 197n2
Kepler, Johannes, 175, 196n1
Keulartz, Jozef, 9, 13n15, 13n17, 42n2, 50n114, 168n2, 171n93
Kierkegaard, Søren, 187
Kiilakoski, Tomi, 129n71
killing in warfare. *See* war
Kitching, Gavin, 43n31
Klein, Jacob, 196n1, 197n2
Kojève, Alexandre, 131n130, 131n131
Korthals, Michiel, 42n2, 168n2
Kuhn, Thomas S., 2, 126n25, 196n1
Kundera, Milan, 129n93
Kurzweil, Raymond, 46n70, 88n88

labor movement, 20, 190, 199n31
LaFleur, William, 45n52
LaFollette, Hugh, 172n116
language game, 33–34
Latour, Bruno, 2, 13n5, 46n76, 90n123, 126n18, 198n24
Law, John, 2
Laws, scientific, 57, 80, 95–96, 98, 125n9, 125n11, 129n79, 145, 147
Lebensform. *See* form of life
legislation, 3, 25, 167, 185, 190
Leibniz, Gottfried Wilhelm, 63, 87n58
life: organic, 59–61; problem of, 58–59; as a theme of ethics, 18–20
leisure, 103–101, 103–104, 110, 132n138
Lenk, Hans, 13n14
Levinas, Emmanuel, 52, 118–119, 131n128, 197n11
Levy, David, 83n2, 83n8, 84n9, 85n31
Lewis, C. S., 91n125, 193, 199n39
liability, 27, 70
Light, Andrew, 13n17, 43n25
Lindberg, Susanna, 86n38, 86n48
Locke, John, 138, 148, 175
logos, 4, 55, 57, 85n30, 142, 159, 164, 166
Lovins, Amory B., 44n36
Lovins, L. Hunter, 44n36
Lucena, Juan, 42n14

Machiavelli, Nicholo, 175, 194, 197n3
MacIntyre, Alasdair, 48n96, 198n22, 205, 206n1
Maclean, Norman, 125n5, 129n82
magic, 80
making. *See poiēsis*

Marcuse, Herbert, 2, 45n54, 46n69, 49n113, 50n117, 83n2, 150
Margolis, Joseph, 168n18
Marjanen, Antti, 84n17, 87n61
Martin, Brian, 44n36
Martin, Mike W., 42n3, 43n17
Martinez-Alier, J., 43n35
Marx, Karl, 2, 12n1, 45n54, 128n68, 140, 152, 168n17
Marx, Leo, 45n51
Marxism, 39, 68–69, 107, 150, 152, 153
material culture, 107, 108, 112, 124, 148, 181, 182, 184, 188
materialism, 55, 57, 62, 88n69
McDermott, John, 169n47
McDonald, David, 131n126
McGinn, Robert, 45n57, 46n59
McLuhan, Marshall, 46n67
means and ends. *See* ends
medicine, 4, 17, 20, 28, 35, 49n106, 90n111, 140, 196
medical ethics. *See* ethics, bioethics; ethics, medical
Merton, Robert K., 2, 12n3
Meskus, Mianna, 198n24–199n27
metabolism, 58–60, 77, 86n43
metaphysics, 32, 37, 145, 179. *See also* Hans Jonas on metaphysics
Mitcham, Carl, 9; defining technology, 4; on engineer vs. humanities philosophy of etchnology, 40; on Hickman's notion of technology, 142; on the relationship between science and technology, 127n35; on willing technology, 192, 199n38
modernity, 67, 80, 90n123, 175–176, 197n3; Max Weber on, 179–180, 203; and metaphysics, 62; and will, 191
modernization, 3, 10, 16, 54, 106, 121
monism (metaphysical position), 55, 86n43
Moor, James H., 42n6
mortality, 36, 60, 61, 73, 89n103. *See also* death
Mulhall, Stephen, 47n90
Mumford, Lewis, 2, 44n45, 50n117, 81
Munda, G., 43n35
Müller, Max, 94
Murdoch, Iris, 173n142

Nash, Roderick Frazier, 43n24–43n25
National Advisory Board on Research Ethics (Finland), 42n9
Nietzsche, Friedrich, 37, 47n78, 49n107, 62, 67, 68, 88n84, 88n87, 194
nihil. *See* nothing; nothingness
nihilism, 49n106, 204; inherent in the modern worldview, 8; Heidegger on, 37–39, 128n64; Jonas on, 55–59, 66, 74, 77, 85n28, 202
Nissenbaum, Helen, 12n2, 13n16, 42n11, 44n42
Nordmann, Alfred, 47n85, 47n87, 125n1
normativity (esp. in ethics), 6, 16, 32, 40–41, 45n54, 46n76, 76, 163, 184, 189, 201
nothing, nothingness, 49n106, 56, 63, 66. *See also* nihilism
nuclear energy, 16, 21, 23, 44n36, 129n72; and polical systems, 44n47
nuclear holocaust, 53
nuclear weapons, 23, 27

The Odyssey, 131n128
Office of Technology Assessment (OTA), 21, 190
ontology, 49n105, 55, 58–61, 63, 70, 77, 86n41, 157, 179, 202; ontological difference, 36; temporal, 67; torah, 197n4
Ortega y Gasset, Jos é, 2, 45n54, 50n117, 132n141
Orwell, George, 128n56
other, Other, otherness: Borgmann on, 98, 110, 116, 121, 193; Jonas on, 70; Levinas on, 118–119, 131n126, 131n128, 197n11
"ought". *See* normativity
Overman (*Übermensch*, Nietzsche's notion), 68, 69, 88n87. *See also* posthuman condition; transhumanism
Ozawa, Hiroshi, 132n146

parenthood, 67–68, 71, 72, 78
Passmore, John, 47n80
Peirce, Charles S., 137–138, 145, 171n114
personhood, 73, 88n78, 89n98, 118–119
philosophy: ancient, 40, 41, 185; analytic, 32, 140, 149; continental, 36, 41, 117,

176; of science, 31, 40, 96, 116–117, 125n11, 126n18, 144–145; of technology, 2, 8, 9, 38, 39–41, 50n118, 81, 95–97, 119–120, 140, 191, 201, 204, 206. *See also* pragmatism
Pinch, Trevor, 2, 90n109, 126n18
Pippin, Robert B., 13n13
Plato, 62, 87n60, 163, 185, 187, 200n42
poiēsis, 15, 29, 31, 144, 167, 182, 203–204. *See praxis*
posthuman condition, 29
Power, Thomas Michael, 127n51
practices. *See* focal things and practices; habituation
pragmatism, 136–137, 140–141; and ethics, 157–159; on facts and values, 156–157; history of, 137–139; and nature, 147; and other philosophical traditions, 149–151; and philosophy of technology, 2, 8, 11; on practice and theory, 153; and science, 181. *See also* Dewey, John; Hickman, Larry A.; Peirce, Charles Sanders
praxis, 15, 29, 31, 39, 117, 140, 146, 167, 182, 203–204. *See also poiēsis*
precautionary principle, 8, 31, 69, 196
prenatal screening, 185–187, 198n23–199n27
principlism, 35, 48n97, 198n20
Pritchard, Michael, 42n13, 42n15, 43n19
production. *See poiēsis*
Prometheus Bound, 89n103
Providence, 55
Prusak, Bernard G., 89n98
Pruuki, Heli, 198n24, 199n26, 199n27
"public," pragmatist notion of, 148

Rabins, Michael J., 42n13
Rahner, Karl, 94
Raittila, Hannu, 136, 168n1
Rathenau Institute, 21, 190
Ravetz, J. R., 43n35
Rawls, John, 7, 39, 77, 94, 113, 148, 205
reading, 170n84
realism: focal, 111; scientific, 31, 95–96, 116–97, 130n111, 126n18, 144–145, 151
recognition (*Anerkennung*), 119, 131n130
releasement. *See Gelassenheit*
religion, 5, 32, 35, 37–38, 46n76, 49n106, 155, 160, 196; and language, 33; as a possible grounding for ethics, 53, 62, 74, 90n108; and technology, 80, 159–160
responsibility, 3, 8, 27–28, 31; attribution of, 46n67; imperative of, 67–73, 75–77, 202; historical background, 27; ontological ground of, 63, 66–67, 192; parental, 67. *See also* liability; parenthood
right, rights, 18, 26, 31, 73, 148, 187, 194
risk, 29, 201; definition of, 43n34; risk assessment, 22, 43n35, 197n6
Rothstein, Bo, 43n30
Rudolph, Kurt, 84n17
running, 110, 120

Safranski, Rüdiger, 197n9
Sartre, Jean-Paul, 85n29
Scheler, Max, 2, 50n117, 143
Schermer, Maartje, 42n2, 168n2
Schopenhauer, Arthur, 79, 192, 195, 199n35, 200n41
Schot, Johan W., 44n44
Schütze, Christian, 83n8
science, 54, 75; and technology, 4, 5, 95–97, 127n35, 153, 156; Levinas on, 119; objects in Science, 116–117; scientific rationality, 31. *See also* realism, scientific; social science; STS; technoscience
science and technology policy, 3, 46n75, 190
Science and Technology Policy Council of Finland, 46n75
scientific realism. *See* realism, scientific
scientism, 169n38
Sclove, Richard, 44n49
self-realization, 25, 45n52, 100, 107
Shrader-Frechette, Kristin S., 21, 43n35, 44n42, 197n6
Snow, C. P., 2, 96
social constructivism (view of technology), 2, 9, 46n76, 126n18, 164, 150
social science, 39, 40, 57–58
Soho, Takuan, 132n150
Son, Wha-Chul, 9, 13n17
sophists, 48n99

Spengler, Ostwald, 2, 91n125
Spier, Raymond E., 42n7
statesman, 72
Stern, David G., 47n87–47n89, 168n16
Stoicism, 55, 77, 96, 197n4
Strauss, Leo, 39, 49n113, 197n3
Strong, David, 43n25, 125n5, 126n20, 126n23, 126n24, 129n79, 130n101, 130n114, 132n143
STS (Science, Technology, and Society / Science and Technology Studies), 2, 8, 9, 40–41, 46n76, 50n115, 81, 119, 126n18, 187, 188, 201, 206
summum bonum (highest good), 149, 68, 196
Suzuki, Daisetz Teitaro, 132n147, 132n149–132n150
Swierstra, Tsjalling, 9, 13n15, 13n17, 42n2, 168n2
system, 4–5, 35, 102–103, 109, 151

Talosela, Minttu, 21–22
Taylor, Charles, 121, 131n130, 132n142
tea ceremony. *See chādo*
technē, 4, 142, 153. *See also poi*; *ēsis*; technique; technology
technique, 4, 142–143, 166
La Technique (Jacques Ellul's concept), 96, 152
technological determinism, 23, 44n45, 96, 152–153
technological progress, 24–25, 34, 45n51, 69–70, 75–76, 79, 100–101
technological vulnerability, 21, 44n36
technology: autonomy of, 23, 44n45, 152, 170n56, 170n82; the "background" of, 103, 106, 177; concept of, 4–6, 96–97, 119–120, 141–145; enabling social progress, 24–25, 100, 131n131; the "foreground" of, 103–104, 106, 107, 128n55; modern 4, 15, 27, 38, 93, 95, 99–100, 199n38; naturalization of, 141–145; promise of, 25, 100, 101, 109. *See also* Borgmann, Albert; the device paradigm; disburdenment; Hickman, Larry; history; science and technology; technoscience, and human will
Technology Assessment (TA), 21–22; Constructive, 22, 190
technoscience, 4, 13n7, 24, 80, 176; Hickman on, 144–145, 150, 154–155, 157, 161, 164; and human will, 191–193
Teknologirådet, 21, 190
television, 1, 101, 102, 111, 117–118, 119, 127n42, 148
temporality, 38, 56–57, 66–67, 112, 153, 168n15, 188, 196
theology, 27, 35, 80, 87n57, 200n40; Catholic, 27, 117; Judaic, 10, 53; Protestant, 27, 83n5, 85n26, 132n150; of technology, 80
theory (*theoria*), 29–31, 144, 167, 203–204
things. *See* focal things and practices
Thomas Aquinas, 197n4
Thompson, Caleb, 197n7
Thompson, Paul T., 9, 132n138, 166
Thoreau, Henry David, 18, 94, 110, 195
Three Mile Island power plant accident, 16
Tolstoy, Leo, 181, 183, 197n7, 200n41
Torah ontology, 197n4
transhumanism, 29, 69. *See also* Bacon, Francis; overman; posthuman
truth, 34, 117; Gandhi on, 194, 200n43; Heidegger on, 5, 38; Levinas on, 118, 197n11; in metaphysics, 61–62, 117
Tuchman, Barbara, 170n56
Turgenev, Ivan, 49n106
two cultures, 7, 12n3, 96, 98

utopianism, 68–69, 75–76, 88n88

values, 99, 157, 158, 161, 176–181, 204; as functions of the will, 79, 200n43; Heidegger's critique of the concept of value, 37–38, 128n64; Jonas on, 54, 57, 61; in nature, 65, 67; within technology, 106. *See also* fact-value distinction; nihilism
Vanderburg, Willem H., 43n22
Verbeek, Peter-Paul, 9, 13n15, 13n17, 132n140, 168n3, 168n20
virtue, 48n99, 80, 87n52, 101, 110. *See also* ethics, virtue; excellence; habituation
Vlastos, Gregory, 198n14
Vogel, Lawrence, 66, 74, 75, 87n57, 88n69
volition, 4, 64, 79, 192, 200n45, 203

voluntary simplicity, 195
Von Wright, Georg Henrik, 47n91, 77, 90n124–91n125, 193, 199n39

Waldstein, Michael, 84n17
Walking, 63, 64, 110, 129n82, 143
war, 148; killing in war, 15
way of life, 120; Aristotle on, 195; in modern societies, 25, 164; "post-technological," 196
weakness of will: See *acrasia*
Weber, Max, 179, 181, 197n7, 203
well-being, 17, 25, 98, 100, 189. *See also* happiness
welfare state, 43n30
Welte, Bernhard, 94
whistle-blowing, 17, 43n17–43n18
White, Lynn, Jr., 90n122
Wiese, Christian, 85n31
Wilderness, 94, 98, 110, 114, 116, 122
will, willing, 79, 144, 192–195, 199n35, 200n43, 200n45, 203, 204
Wilshire, Bruce, 125n5
Winner, Langdon, 9, 43n18, 43n22, 44n43, 44n45, 44n46, 44n49, 46n76, 48n92, 88n88, 128n66, 129n72, 150, 151, 152, 153, 164, 170n56, 182
Wittgenstein, Ludwig, 5, 6, 13n11, 32, 140, 197n7; "A Lecture on Ethics," 33; *Culture and Value*, 34, 47n91; *Philosophical Investigations*, 33–34, 47n87–47n89, 165; *Tractatus Logico-Philosophicus*, 33, 47n85. *See also* form of life; language game
Wolf, Clark, 46n64
Wolin, Richard, 49n113, 83n2, 83n4–83n6, 83n8–84n9, 88n78, 90n115
work, technologically structured. *See* technology, the "background" of
the world, 5–6, 7, 55–58, 63, 75, 80, 87n61
Woolgar, Steve, 2
Wynne, Brian, 172n131

Zahavi, Dan, 130n117
Zen Buddhism, 123, 132n150

www.ingramcontent.com/pod-product-compliance
Lightning Source LLC
Chambersburg PA
CBHW051252200926
51612CB00001B/25
* 9 7 8 0 7 3 9 1 6 8 6 4 6 *